OUR NEIGHBOR STARS

INCLUDING BROWN DWARFS

by
Thomas Wm. Hamilton

Strategic Book Publishing and Rights Co.

Strategic Book Publishing and Rights Co.
12620 FM 1960, Suite A4-507
Houston, TX 77065
www.sbpra.com

ISBN: 978-1-61897-132-6

Book Design: Suzanne Kelly

Other Books by Thomas Wm. Hamilton

Time for Patriots: The 21st Century Confronts Bunker Hill—And After!

Useful Star Names, with Nebulas and Other Celestial Features

TABLE OF CONTENTS

PART 1

SO WHO IS CLOSE?

Prior to 1838 no one knew the distance to any star except the Sun. Estimates of the Sun's distance went back to Aristotle, or even earlier, and while some of those estimates were way off, by the early Nineteenth Century astronomers knew the Sun was over 80 million miles away. But for other stars all they could do was guess.

In the 1830s telescopes had improved to the point where professional astronomers began to hope they could actually measure the distance to nearby stars. They knew it would be difficult, so they hoped to start with nearby stars, just to make it a bit easier. But if you did not already know which stars were close to us, how could you pick out which ones to start with?

One approach was to assume (as even Aristotle had 24 centuries ago) that the brightest stars were the closest. The brightest of all stars (after the Sun) is Sirius, a star whose very name meant bright or blazing in Ancient Greek. A more technical approach said that stars are moving. This is called proper motion, and the higher the proper motion, presumably the closer the star. You can see this effect when riding in a car. Distant objects move across your field of view more slowly than do nearby ones. Of course, with stars, both the Sun and the star will be moving, but the idea seemed logical.

The fastest moving star known in the 1830s was a visible but moderately dim star called 61 Cygni, located in one of the wings of the constellation of Cygnus the swan. The German astronomer Friedrich Georg Wilhelm Bessel (1784-1846) picked it to be the first to measure. The Baltic astronomer Friedrich Struve (1793-1864) (founder of a famous dynasty of astronomers) opted for a very bright star, Vega, since Sirius, the brightest, is so far south in the sky as to be badly placed for measuring from Struve's location at Pulkovo Observatory near St. Petersburg, Russia. The second brightest star, Canopus, is so far south that it is never visible from Pulkovo. Thomas J. A. Henderson (1798-1844), the first Astronomer Royal of Scotland, had not yet reached that honor, and was working at Table Mountain Observatory in South Africa. Cleverly, he chose a star high on both lists, of fast movers and of very bright stars. This was the star known as Alpha Centauri.

Henderson completed his work first, but disbelieved his results, which showed Alpha Centauri to be slightly over 3 lightyears away. He doubted it could really be that far. So Bessel published first, coming up with a distance for 61 Cygni that was even greater, over ten lightyears. This convinced Henderson to publish. Struve followed shortly after.

Each of the three came up with figures a bit under the correct distances for their respective stars, but close enough to the modern results to prove they had in fact measured the distances within a reasonable margin of error. Possibly their consistency in being a bit low resulted from bias against believing the truly enormous distances involved.

Each used the same method, which is still used today, although most accurately by the European Space Agency's satellite Hipparcos rather than astronomers on the ground squinting through a telescope. (Actually, professional astronomers today *never* look through telescopes. In fact, it is impossible to look through modern large telescopes, they only photograph or make other records.)

So how was the measurement made? I'll start with an example. Hold your arm out straight, and raise a finger. Close one eye, and note where the finger appears against the background. Now, without moving your arm or finger, switch eyes. The finger will appear to shift its position against the background. This is due to the difference in the locations of your two eyes. It is how humans are able to judge distances. For stars, the baseline difference between your eyes is a bit inadequate for the job, but we have a much larger baseline available. Earth averages 92,955,807.27 miles (or 149,597,870.7 kilometers) from the Sun's center, a value known as the Astronomical Unit, or AU. So get the location of a star against the background, wait six months, and do it again. This gives a baseline of nearly 186 million miles or 300 million kilometers.

Even with such a large baseline the stars are all so far away that the shift is very small. In fact, observers from Aristotle (384 BC - 322 BC) to Tycho Brahe (1546 - 1601) used the fact that no shift was detectable to "prove" that the Earth stood still. The shift is known as the star's parallax, and even for the closest star is well under one second of arc, where a full circle is 360 degrees, each subdivided into 60 minutes of arc, each of which is subdivided into 60 seconds of arc. So one second of arc is just one part in 1,296,000 of a circle. I don't know how good Aristotle's eyesight was, but no one can see detail that fine!

With the introduction of photography to the problem late in the Nineteenth Century, results became easier and more accurate. Yet the angles were so small that distances could be measured only out to a few hundred lightyears, and with a generous margin of error. Quoted distances were generally accurate to only +/- 15 or 20 percent, and were essentially meaningless after about 500 lightyears. The work was slow and tedious, and by 1990 fewer than ten thousand stars had any sort of distance measured.

The European Space Agency's Hipparcos satellite changed all that. It was able to accurately measure distances to 1600 lightyears, and determined it for over 100,000 stars.

Astronomers for several decades had felt that they might have an almost complete census of the stars within 16 lightyears of the Sun. That every few years another star would be added to that list suggested that they may have been somewhat optimistic. But having even a pretty good census meant that astronomers would get a feel for how common each type of star was.

STARS COME DIFFERENT TYPES?

Anyone looking at the stars on a clear night can see that they have different colors. Betelgeuse in the winter sky and Antares in the summer are distinctly reddish, a few have a blue tint. By the Nineteenth Century astronomers recognized that this meant they were different temperatures, and a Jesuit astronomer, Father Angelo Secchi (1818-1878), organized the different types according to their temperatures/colors. Hottest were the blue stars, with temperatures ranging down from there through white, yellow, orange, and red. Letters had been assigned to the different colors before Secchi, and were so embedded in the literature that he did not consider changing them. The result is that today, with the addition of several letters to cover types unknown to Secchi, in descending order of temperature star types are W, O, B, A, F, G, K, M, R, N, S, L, T, and Y. The last three are for brown dwarfs rather than real stars. Generations of students have gasped in dismay at the thought of having to remember that sequence, until learning the unforgettable mnemonic, "wow, oh be a fine girl, kiss me right now sweetheart, lovingly, tenderly". Who said astronomers aren't fun guys? (Don't blame Fr. Secchi for that mnemonic, it supposedly was invented at Harvard.)

Early in the Twentieth Century the Danish astronomer Ejnar Hertzsprung (1873-1967) and American astronomer Henry Norris Russell (1877-1957) collaborated on creating a chart or graph on which they plotted stars according to their temperatures and intrinsic brightness. Of course, the apparent brightness in the night sky was easy to see, but how bright were the stars really? For that you needed the distance. Distances for a few hundred stars were known by then. Hertzsprung and Russell found that most stars on their chart fell on a curve vaguely resembling a shallow backwards N. Since this included most stars, it became known as the Main Sequence. Other stars fell on a branch that included giant stars, so it was the Giant Branch. A group of very small and hot stars comprised the white dwarfs. Today that graph is known in honor of its creators as the HR Diagram, and it provides a key to understanding the nature of individual stars.

The letters (O, B, A, etc) get subdivided by finer gradations of temperature using numbers from 0 to 9, with 9 the coolest and 0 the hottest. Thus G9 is just warmer than K0. The Sun is ranked as a G2, with a surface temperature of about 9950F, or 5800K (the interior of sunspots is cooler). Stars on the Main Sequence are indicated by the roman numeral V, so the Sun's type is G2V. Giant Branch stars and some others have a complex set of types, with roman numerals I, II, etc. It is now known that Giant Branch stars are approaching the ends of their lives. White dwarfs are basically the burned out hulks of stars. In the very remote future the Sun will evolve into a giant red star, and then slowly deflate into a white dwarf.

Its mass will not change significantly, so its density will drop very low, and then shoot up to an amazing amount.

Perhaps somewhat contrary to expectations, the larger and more massive a star is when it forms, the shorter its lifespan. Red dwarfs with just 15% of the Sun's mass are good for 300 or 400 billion years. The Sun is already 4.35 billion years old, and is good for about 1.2 billion years before starting to swell up. A star such as Sirius, with more than twice the Sun's mass, has a total life span of little more than a billion years.

Astronomers found that the most common type of star was the red dwarf, even though with their small size and limited light output they were dim and hard to find. Bright stars are comparatively rare, and our galaxy of over 200 billion stars may have only a couple dozen stars of the W type.

Stellar types get a lot more complicated, depending on the ratio of the amounts of various elements to hydrogen, the speed of the star's rotation (this affects how well and how fast it mixes its contents, as well as its shape), the presence and distance of stellar companions, emissions, and much more. But this is a work on what stars are near us, not astrophysics, so let's lay that topic on the table.

WHAT DISTANCE?

On Earth Americans measure distance in miles, and just about everyone else on the planet measures it in kilometers. Knowing that the Sun is 93,000,000 miles, or 149,500,000 kilometers away hints that the stars' distances are going to be so far that writing it down will lead to a bad case of writers' cramp, waste of time writing, and take up too much paper and ink. Efficiency is called for.

Light moves in a vacuum at its maximum possible speed, 186,282.3959 miles per second, or 299,792.458 kilometers per second. After astronomers started finding stars were an enormous distance away, they invented the lightyear, the distance light travels in one year. You could figure out the exact amount easily enough, just multiply the miles per second by the number of seconds in a year ($60 \times 60 \times 24 \times 365.2422$), but it has already been done for you. A lightyear is 5,878,625,373,183.608 miles, or 9,460,730,472,580.8 kilometers. Earth's distance from the Sun is called the Astronomical Unit (AU), and is usually used for distances within the Solar System. A lightyear is also 63,241.1 Astronomical Units.

MAGNITUDES

Around 160 BC the Greek astronomer Hipparchus (190 BC - 120 BC) invented the magnitude scale for discussing how bright stars are. He declared the twenty brightest stars not counting the Sun (this includes three in this list) as being the first magnitude. The dimmest stars visible were lumped together as sixth magnitude. Note that the larger the number the dimmer the star. Those in between were second, third, fourth, and fifth magnitude. This served well enough until the invention of the telescope revealed a horde of stars fainter than sixth magnitude. Things drifted with no definitive resolution until the mid Nineteenth Century, when Benjamin A. Gould (1824 - 1896) created a mathematical formula for magnitudes. Each full magnitude would be the fifth root of 100 (about 2.512 times) brighter or dimmer than the next full magnitude. The way this worked out three of Hipparchus's stars wound up with negative magnitudes. A difference of five magnitudes is exactly 100 times brighter or dimmer. And Hipparchus got the European Space Agency spacecraft Hipparcos named for him.

Apparent magnitude refers to how bright the star looks in our sky. The brightest star is the Sun, with an apparent magnitude of -27. The next brightest star is Sirius with an apparent magnitude of -1.46. A difference of exactly 25 magnitudes would mean that the Sun looks 100^5 times brighter (10,000,000,000) than Sirius appears to be. Absolute magnitude refers to how bright a star would look if placed at a standard distance. The Sun's absolute magnitude of about 4.6 makes it a very ordinary star, although few stars in this list have a brighter absolute magnitude.

ACCURACY

When I worked on the Apollo Project the engineers were extremely upset to learn that the Moon's distance was then known only to an accuracy of one mile. As the only astronomer around I heard the brunt of the complaints. My response was that if one mile made a difference when the astronauts had to go 230,000 miles, their tolerances were too tight. Today, with laser reflectors on the Moon, its distance is known within one centimeter (0.4 inch), an accuracy which has permitted verification of the theory of continental drift on Earth.

We generally do not have the distances of most other Solar System objects to an accuracy much better than a mile today, and for many minor objects such as asteroids, accuracies may be in the range of a hundred miles. Outside the Solar System the distances of the handful of stars within ten lightyears probably are accurate to the first four significant digits, but as we move further outward the accuracy drops. A star listed as 17.00 lightyears may actually be somewhere in the range of 16.85 to 17.15 lightyears. Thus the order of stars on this list may change as one is found to be slightly more distant and another slightly closer.

There is also the question of completeness. Brown dwarfs in particular are so faint as to be hard to find, and dimmer red dwarfs and white dwarfs may have been overlooked. There is a high probability that in the future from five to ten objects not yet on this list will be found to fall within a list of the closest. As evidence of this, entry #13 below was only reported discovered in August of 2011. The list of 100 stars is probably fairly complete, but the list of thirteen nearest brown dwarfs may actually triple over the next few years.

WHATCHMACALLIT?

Few stars actually have been given names. (For more on that, see my earlier work, *Useful Star Names,* Strategic Book Publishing, 2011, ISBN 978-1-61204-614-3.) Most are referred to by catalog numbers. The problem here is that there are many catalogs. Bayer started the use of Greek letters. Flamsteed used numbers, assigned going from east to west. Gliese numbered nearby stars. Members of the Struve family invented designations for variable stars. The Nineteenth Century had the Henry Draper Catalogue and the Bonner Durchmusterung, plus its annex, the Cape or Cordoba Durchmusterung. The Twentieth Century had the Smithsonian Astrophysical Observatory Catalog. Our era has the Hipparcos Catalog. And any one star can appear in all of these, plus a few more I haven't mentioned. In listing the nearest stars, I have tried to include most of the catalog designations people are likely to be using, so the stars may be found more easily. Check the alphabetical index near the end if you are seeking a particular star, as this is where I have listed a number of different designations for most stars.

PART 2

THE CLOSEST STARS

1. The closest star is the *Sun*. It has eight known planets, half a dozen dwarf planets, and hordes of comets, asteroids, and moons of planets. It rotates fastest at the equator, about once every 25 days, and slower towards the poles, about once every 32 days. The Sun fuses hydrogen into helium in a process in which four protons combine in a series of steps that turns two into neutrons and leaves the result as a helium nucleus (two protons, two neutrons). This was first worked out around 1950 by a University of Michigan astronomer, Subrahmanyan Chandrasekhar (1910 - 1995). For this he won a Nobel Prize in physics, and the Chandra Space Telescope is named for him. Spacecraft designed specifically to study the Sun include the OSO (Orbitting Solar Observatory) series launched by NASA in the 1960s, two Helios spacecraft from Germany, STEREO, SOHO, and others.

The mass of the Sun is the standard by which other stars are measured, so its mass is stated as 1. It can also be expressed as 330,000 times the mass of the Earth. The temperature in the undisturbed solar surface is 9950F (about 5800K), but is cooler inside sunspots, down to about 8000F. The visible surface of the Sun is called the photosphere. Above the photosphere is a very thin layer called the chromosphere. Above that is the Sun's atmosphere, called the corona. The corona is normally only visible during solar eclipses, when the body of the Sun is hidden and the corona is visible around the edge of the Moon. The size and shape of the corona changes with the Sun's sunspot cycle, ranging from half a million to two million miles out from the Sun's surface.

While we cannot see below the photosphere, there is good evidence that the Sun has a core about 80,000 miles across where the energy is produced, with two layers, the radiative (NOT radioactive please, energy radiates outward) and convective, between the core and photosphere. The Sun does *not* burn, a chemical process, as it is too hot for any chemicals to exist.

Sunspots are places on the photosphere slightly less hot than the rest of the surface. Frequently we find sunspots associated with flares, violent surface explosions on the Sun which shoot streams of particles (neutrons, protons, electrons) out into space along with bubbles of magnetic energy. If these are aimed at Earth, we will often see auroras, which are caused by these streams impacting molecules in Earth's upper atmosphere and reacting with the planet's magnetic field. Other surface features on the Sun include faculae (hot clouds of gas) and granulation, a mottling effect caused by hot bubbles rising from inside to the surface. These granules start out bright and gradually dim as they cool, to sink back inside. Spicules are spikes of flame extending upward from the photosphere for a few hundred miles. Prominences are giant eruptions of flame which can extend into space for half a million miles or more, and are usually found in association with sunspots and flares. Sunspots, flares and prominences fluctuate in numbers over an eleven year cycle. This was first noticed in connection with numbers of sunspots around 1843 by a German astronomer, Samuel H. Schwabe (1789-1875), and so is called the sunspot cycle. The Sun's magnetic field actually reverses itself every eleven years, so a complete cycle would be 22 years.

Absolute magnitudes are how bright a star would be if the star were 32.56 lightyears away. The Sun, which has an apparent magnitude of -27 because it is so close to us, has an absolute magnitude of +4.6, rather dim, but visible in a clear dark sky if we were to ever be that far from it. If that distance seems a trifle odd, it is exactly ten parsecs, an alternative measure of stellar distances.

The Sun is believed to be about 4.35 billion years old, and to be in a fairly circular orbit around the center of our galaxy, which is in the direction of the constellation Sagittarius 28,000 lightyears away. The orbit is slightly tilted with respect to the central plane of our galaxy, so we are roughly 110 lightyears off plane currently.

2. *Proxima Centauri* (also known as *Alpha Centauri C, HIP 70890* [from the Hipparcos catalog], *V645 Centauri* [the 645th variable star found in the constellation of Centaurus], *GJ 551* [the 551st star in a list of nearby stars compiled by the German astronomers Wilhelm Gliese {1915-1993} and Hartmut Jahreiss {1942-}], and other designations) is a dim red dwarf of class M5.5Ve discovered in 1915 from the Union Observatory in South Africa by the Scottish astronomer Robert T. A. Innes (1861-1933). At a distance of 4.243 lightyears, it is 270,000 times further away than the Sun. The distance was first measured by American astronomer Harold Alden (1890-1964) in 1928. Its diameter is only 14.5% that of the Sun, about 126,000 miles, and the mass is 12.3% of the Sun. Most of Proxima Centauri's light output (85%) is in the infrared, not visible to the human eye, making it even dimmer. The apparent magnitude is +11.05 (five magnitudes too dim to see with just your eyes). The absolute magnitude is 15.49, very dim. This gives it a luminosity only 0.17% as bright as the Sun.

In 1951 Harlow Shapley (1885-1972) discovered Proxima has flares that increase its brightness by up to 8%. Its normal surface temperature is just 3042F, although inside flares it will increase a few thousand degrees briefly. The dimness is a consequence of the low temperature and small size of this star. Proxima is known to have strong convection currents, which keep its interior well stirred. This has the effect of prolonging its life, since it will get to use up nearly all its hydrogen. So where the Sun has merely 1.2 billion years before starting to expand into a red giant, and not much over 2 billion years before winding up a white dwarf, Proxima may hang around pretty much the way it looks today for another 400 billion years on top of the 4.85 billion it has already existed.

No planets have been discovered around Proxima, but if it has any, the so-called life zone, or Goldilocks zone, where water would form a liquid on the surface, would be just 0.023 to 0.054 astronomical units (2.14 to 5.02 million miles) from the star. At such distances a planet's orbital period around Proxima would be just 3.6 to 14 days. This close even its feeble gravity field would probably have captured a planet into always keeping one face to the star, the way the Moon always has one face towards Earth.

Proxima is moving through space, as do all stars. This is called the star's proper motion (across our line of sight) and radial velocity (motion towards or away from us). For Proxima, the radial velocity is -22.4 kilometers per second, the minus sign meaning it is approaching the Solar System. Its closest approach will occur in 27,400 years, when it will be 2.90 lightyears away, and still too faint to see. That's a big miss, so don't worry.

Proxima is probably in an enormous elliptical orbit around Alpha Centauri, taking close to half a million years to complete one orbit. Currently it is 15,000 astronomical units from Alpha Centauri (0.237

lightyear or 140 billion miles), near the maximum distance in its elliptical orbit. That far away it is merely fifth magnitude (+5.15 to be precise) from any possible planets around Alpha Centauri. Alpha Centauri would be magnitude -6.26 from Proxima. That's about 1.5 magnitudes brighter than Venus can appear in our sky, and bright enough to see in the daytime.

Because of the flares, it is very unlikely that life resembling anything on Earth could exist, and Earth life would have trouble surviving over the long term.

Proxima Centauri is located at Right Ascension 14 hours 29 minutes 42.9 seconds, Declination -62 degrees 40' 46.1" in the constellation of Centaurus.

3 & 4. *Alpha Centauri* is really two stars, A and B. The brighter star is A, with an apparent magnitude of -0.1. This pair is 4.365 lightyears from the Solar System, or about 25,800,000,000,000 miles. Now you understand what I meant about writer's cramp and taking up too much space as an explanation for why the lightyear was invented.

Alpha Centauri A is about ten percent more massive than the Sun, with a diameter 22.7% bigger (1,060,000 miles), and a luminosity 1.52 times that of the Sun. It rotates once every 22 days, a bit faster than the Sun. Both the A and B stars are about 4.85 billion years old, around 400 million years older than the Sun. Alpha Centauri is so far south in the sky that any point in the United States north of Cape Canaveral cannot see it. That Alpha is composed of two stars was discovered by the French Jesuit astronomer Fr. J. Richaud (1633-1693) in 1689, observing from India.

Alpha Centauri's orbit in the galaxy is tilted by about 10 degrees, rather more than most stars. Its orbit will bring it closest to the Solar System in 27,900 years, when it will be 3.26 lightyears away in the constellation of Hydra. Alpha A has not yet been found to have any planets, which probably means that it does not have any giant planets similar to Jupiter, although smaller planets in the size range of Earth are possible.

Alpha Centauri A is classified as G2V, same as the Sun, although slightly older and larger. As its name tells us, it is in the constellation Centaurus.

Alpha Centauri B is smaller, 86% the diameter of the Sun (743,000 miles), and 90.7% of its mass. The star rotates slowly, once in 41 days. Remarkably, it is a source of X-rays, emitting more of them than the Sun or its two companions. Alpha B goes around A in an elliptical orbit taking 79.94 years to complete. Its distance from A varies from 11.2 AU (1,025,000,000 miles) to 35.6 AU (well over 3 billion miles). The last time they were closest (a position called periastron) was August 1955, and next will be July 2035. Most distant (called apastron) was May 1995, and next will be in 2075.

As with its companion, no planets have yet been found. Any planet in orbit around B at a distance giving it a climate similar to Earth's would have to be about 70 million miles from the star. Alpha Centauri A would be very bright (-21.92 at closest), but far enough off and dim enough not to create problems for plant or animal life unless they need dark nights. However, at opposition (meaning A would be up at midnight) you would be able to read a newspaper by A's light. From planets around A, B would be -20.59 when the stars are closest, -18.11 when they are furthest apart.

Alpha Centauri B is classified as K1V. Its apparent magnitude is 1.33, the absolute magnitude is 5.71, and surface temperature is 5260K.

Our Sun is the closest neighbor to the Alpha Centauri system, with Barnard's Star (see #5) as their next closest neighbor at 6.5 lightyears. There are only seven stars within ten lightyears of this trio.

Alpha Centauri is located at Right Ascension 14 hours 39 minutes 36 seconds, Declination -60 degrees 50' 02".

5. *Barnard's Star* was discovered in 1916 by the American astronomer Edward Emerson Barnard (1857-1923), who noted its fast proper motion. Barnard was working at Lick Observatory in California. The star is located in the constellation Ophiuchus, 5.9 lightyears from us. A red dwarf, its apparent magnitude is 9.54, much too faint to see without optical aid, with an absolute magnitude of 13.24. It has the highest known proper motion, moving across our line of sight at the rate of 10.4 seconds of arc per year. Its motion will bring it nearest the Solar System in the year 11,700 A.D., when it will be 3.8 lightyears away, still too faint to see with naked eyes.

Barnard's Star is one of the slowest rotating stars known, completing one rotation in 130.4 days. Its mass is 17% that of the Sun, while its diameter is 20% that of the Sun (173,000 miles). The surface temperature is a cool (for stars) 3134K, and its luminosity is just 0.0004 that of the Sun. Astronomers believe Barnard's Star is at least nine billion years old, more than twice the age of the Sun.

In the 1950s Peter van de Kamp (1901 - 1995) reported finding evidence of one or two planets with masses close to that of Jupiter around Barnard's Star. This later turned out to be an error caused by renovations to the telescope he was using. No planets are known, and any approaching the mass of Jupiter are extremely unlikely. Because Barnard's Star is so old, it formed at a time when our galaxy had far less of the higher numbered elements, and it therefore is unlikely to have planets with useful minerals, if any planets at all. Despite this, Barnard's Star has been a popular target in science fiction.

There are eleven stars within ten lightyears, with its nearest neighbor being Ross 154 (see #12), 5.5 lightyears away from Barnard's.

Barnard's Star is classified as M4Ve, the e signifying emission lines in its spectrum. In 2000 it was located at Right Ascension 17 hours 57 minutes 49 seconds, Declination +04 degrees 41' 36".

6. *Wolf 359* is another red dwarf, classed as M6.5e. Maximilian Franz Josef Cornelius Wolf, better known as just Max Wolf (1863-1932), created a list of fast moving stars. This star was number 359 on his list, and it has been known as Wolf 359 ever since. It is located in the constellation of Leo, although at apparent magnitude 13.54 is much too faint to see without a telescope. The absolute magnitude is 16.65. It is 7.78 lightyears away.

This star was nearest the Solar System 13,850 years ago, when it was 7.35 lightyears away, and of course was invisible then because no one had telescopes.

Wolf 359 has a mass 9% that of the Sun, and a radius of about 68,000 miles (compare Jupiter, with a radius only a third smaller). Its surface temperature is so low, just 2800K, that it has a rarity among stars, chemical compounds. Most stars, including the Sun, are far too hot for any kind of chemical compounds to exist (that's one reason why we cannot say stars are "burning", since burning is a chemical process). Wolf 359 however has had water, titanium oxide, carbon monoxide, iron hydride, and a few other compounds identified. Unrelated, but also impressive, Wolf 359 has a magnetic field much stronger than the Sun's.

The closest stars to Wolf 359 are Ross 128 (see #16) and Lalande 21185 (see #7), at 3.8 and 4.1 lightyears respectively. There are 18 stars within ten lightyears.

As is common among red dwarfs, Wolf 359 has no radiative layer. Energy is conveyed from the fusion core to the surface only by convection. This has the side effect of keeping the star's contents well mixed. Where the Sun and similar stars will eventually start to evolve onto the giant branch because the core is clogged with helium produced by fusion when about 30% of the hydrogen has been converted to helium, smaller and cooler stars will be able to fuse virtually all their hydrogen. The result is that Wolf 359, believed to be under one billion years old, has a life expectancy of 400 trillion (400,000,000,000,000) years. After just one tenth of one percent of its lifespan, the Sun and our planets will be long forgotten.

Wolf 359 is located at Right Ascension 10 hours 56 minutes 29 seconds, Declination +07 degrees 00' 52".

7. *Lalande 21185* is another red dwarf named for the catalog it appears in. Jerome Lalande (1732-1807) published an important star catalog in 1801 that included this star, located in Ursa Major. Its apparent magnitude is 7.47, with an absolute magnitude of 10.44. It is 8.29 lightyears away. Lalande 21185 is the third brightest red dwarf in our sky.

The first person to measure its distance was Friedrich August Theodor Winnecke (1833-1897) in 1856. As seems to have been common with these early measurements, he made it a bit too close. However, in just 19,900 years it will be a lot closer, about half the current distance, making it just barely visible to the human eye under perfect conditions.

Lalande 21185 has a diameter of about 400,000 miles and a surface temperature of 3383K. Its orbit around the galactic center seems to be inclined a full 90 degrees to the plane of the galaxy.

The question of whether Lalande 21185 has planets has a convoluted history. In 1951 Peter van de Kamp reported that work at Swarthmore College's Sproul Observatory had revealed at least one planet. His former student Sarah Lippincott (1920-) in 1960 confirmed this. In 1974 George Gatewood (1940-) showed that as with Barnard's Star this was an error based on changes in the telescope used. Then in 1996 Gatewood claimed to have found evidence of planets! No one else has been able to confirm this, and the current situation is that if any planets exist, they probably are small, and have yet to be detected.

After Wolf 359, Lalande 21185's nearest neighbor is Ross 128 (see #17) at 6.5 lightyears. There are thirteen stars within ten lightyears.

Lalande 21185 is classified as M2V. It is at Right Ascension 11 hour 03 minutes 20 seconds, Declination +35 degrees 58' 12".

8 & 9. *Sirius* is located in the constellation of Canis Major, 8.58 lightyears from the Solar System. It is actually a binary, or two stars, Sirius A and Sirius B. Sirius A did not have to be discovered, as it is the brightest star in the night sky, at apparent magnitude -1.46. The absolute magnitude is +1.42. Sirius A has a mass 2.02 times that of the Sun, a diameter 1.711 times that of the Sun (1,478,000 miles), and is 25.4 times more luminous than the Sun. It is a class A1V star.

Given its mass and temperature, Sirius A does not use the same method of generating energy that the Sun does. Massive stars use what is called the CNO method. In this a carbon atom picks up an extra proton, and through a series of steps picks up three additional protons while changing first to nitrogen and then to oxygen. The final step has the oxygen emitting an alpha particle (helium nucleus), returning to being a carbon atom. The net effect is the same as in the Sun, four protons turned into a helium nucleus, while the carbon atom is restored unchanged.

In 1854 Bessel noted that Sirius's proper motion was a wavy line, suggesting that a companion star's gravity was influencing the motion. In 1862 the American telescope manufacturer Alvan Graham Clark (1832-1897) was the first to see this star while testing a newly completed telescope. In 1915 Walter S. Adams (1876-1956) got the spectrum of this star, discovering that it was extremely hot, yet very faint at apparent magnitude 8.44. The calculated absolute magnitude was 11.34, giving it a luminosity 2.6% that of the Sun. Since its mass could be calculated from its gravitational effect on Sirius A, it appeared this star had a mass 98% that of the Sun. This perplexed and shocked astronomers. Suggestions included the idea that spectra did not truly represent temperatures, that the star was really very far away and a different, closer star was influencing Sirius, or that something was hiding the star's light. As we now know, they had found the first white dwarf.

Sirius B, sometimes known as *The Pup,* orbits A in 49.94 years. The orbit is fairly elliptical, ranging from 8.1 AU (753 million miles) to 31.5 AU (2.9 billion miles). The stars undoubtedly formed at the same time, but Sirius B was initially the more massive of the two, probably a class B5V star. It went through its evolution rapidly, ending as it is today. Sirius A being less massive will not evolve into a white dwarf for a couple hundred million more years. The life span of both stars is so short that it is unlikely any planets around either could have developed life, and the dying stages of B would have thrown out enormous masses of hot gases which would have probably sterilized any planets around either star.

For a planet around A the white dwarf would range from magnitude -12.75 to -15.70, keeping nights very bright. A planet around the white dwarf would see A ranging from -22.62 to -25.57. The latter figure is less than two magnitudes fainter than the Sun appears in our sky.

Sirius B has a diameter of only 9300 miles, slightly larger than Earth, but with more than 300,000 times as much mass. This gives it an enormous density, and explains why, despite being very hot and close, it is so faint. Each square inch of surface is putting out a lot of light, it just doesn't have that many square inches compared to most stars.

The closest star to Sirius is Procyon (see #21), 5.2 lightyears away, followed by Luyten's Star (see #37) at 5.8 lightyears. There are fifteen stars within ten lightyears.

Sirius is located at Right Ascension 6 hours 45 minutes 09 seconds, Declination -16 degrees 42' 58".

10 & 11. *Luyten 726-8* is another pair of red dwarfs named for their catalog entry. Luyten 726-8 was reported as a single star in 1948 by the Dutch astronomer Willem J. Luyten (1899-1994), who later discovered that it is a binary. Star A has an apparent magnitude of 12.57, and an absolute magnitude of 14.92. Its mass is one tenth that of the Sun, and diameter is one seventh the Sun's (123,400 miles). The luminosity is 0.06% the Sun's. Star A is a flare star classed as M5.6Ve, and is also called *BL Ceti.*

Star B has a 26.5 year orbit around A, ranging from 2.1 AU (295 million miles) to 8.8 AU (818 million miles). Its mass and diameter are the same as its companion's, but the apparent magnitude is 11.99 and absolute magnitude is 15.37. The luminosity is 0.04% of the Sun. As with so many red dwarfs, it is also a flare star, known as *UV Ceti.* It is so typical of flare stars that the general class of red dwarf flare stars are known as UV Ceti types. The surface temperature outside of flares is 2670K.

Luyten 726-8 is 8.73 lightyears from the Solar System in the constellation of Cetus, and is much closer to Tau Ceti, only 2.87 lightyears. It was nearest the Solar System 28,700 years ago, when it was 7.2 lightyears away (and of course still not visible). In 31,500 years it will pass only 0.93 lightyears from Epsilon Eridani, which could stir up the Kuiper Belt of both systems, assuming they have Kuiper Belts. Given the flares, even if one or both stars should have a planet close enough for liquid water, life would have a very hard time existing.

There are sixteen stars within ten lightyears.

Luyten 726-8 is located at Right Ascension 01 hours 39 minutes 02 seconds, Declination -17 degrees 57' 02".

12. In 1924 Frank E. Ross (1874-1960) was appointed Director of Yerkes Observatory, where he had access to thousands of photographic plates shot over many years by Edward Barnard. Ross decided to make new photos of the regions shot by Barnard, and compare the pictures using a device called a blink comparator. (It was a blink comparator which revealed the existence of Pluto a few years later.) Changes between old and new photos permitted Ross to identify several hundred previously unknown variable stars plus many presumably nearby stars from their proper motion. Number *154* on his list of high proper motion stars turned out to be the red dwarf which occupies twelfth place in our list of neighboring stars. It is 9.68 lightyears away, with an apparent magnitude of 10.93. The absolute magnitude is 13.07, but it has flares on a semi-regular basis every two days. This increases its brightness by three to four magnitudes (sixteen to forty times brighter!). The mass is 17% that of the Sun, its diameter is 24% that of the Sun (208,000 miles), and the surface temperature is 3105K, except inside the flares. The luminosity is 0.38% of the Sun. 150,000 years ago Ross 154 was nearest the Solar System, 6.13 lightyears. This star is believed to be under one billion years old, and has not yet settled down, explaining the flare activity.

The flares make it almost certain that no life could exist at Ross 154, even if it should turn out to have planets.

At 5.5 lightyears, Barnard's Star is closest to Ross 154, followed by Lacaille 8760 (see #42) at 7.4 lightyears. There are seventeen stars within ten lightyears.

The star is located at Right Ascension 18 hours 49 minutes 49 seconds, Declination -23 degrees 50' 10", in the constellation of Sagittarius. It is classified as M3.5Ve.

13. *WISE 1541-2250* holds several records. It is the closest known brown dwarf, the coolest known brown dwarf (a surface temperature of just 80F, or 25C), and one of the first brown dwarfs discovered of type Y. It is Y8V. Type Y is characterized by extremely low temperatures, with all sorts of interesting chemicals in its atmosphere. This one is in the constellation of Lyra.

Brown dwarfs do not have enough mass to create the heat and pressure which would permit fusion of hydrogen into helium by any known process. They could briefly fuse deuterium into helium, but deuterium is rare, and this would not last very long. For type Y, if such fusion ever did occur, it was very brief and is long since ended. Any planets in orbit around such a brown dwarf would be frigid. The question has been raised as to whether a brown dwarf could support life. Enough warmth is available for that, and the mix of chemicals present make that barely possible. Unfortunately, extreme turbulence probably occurs, making the development of life forms impossible in brown dwarfs.

WISE 1541-2250 is located at Right Ascension 15 hours 41 minutes, Declination +22 degrees 50'.

So in a sphere with a radius of ten lightyears around us we have a dozen stars and one known brown dwarf. The stars average about 3.5 lightyears apart, but this is skewed by having a triplet and several binaries. Each star has about 4.3 cubic lightyears for itself. Were we located in a cluster these numbers could be much smaller. Actually, Sirius is the nearest member of a minor cluster which spreads out over more than a hundred lightyears.

14. Frank Ross next offers *Ross 248*, also known as *HH Andromedae*, from his 1926 catalog. This M5.5V red dwarf has an apparent magnitude of 12.23 to 12.34 thanks to occasional flares. The absolute magnitude during quiet times is 14.79. It is 10.32 lightyears away. The star has a mass 12% that of the Sun, a diameter 16% that of the Sun (144,000 miles), and a luminosity of 0.2% of the Sun. There seems to be a fairly regular 4.2 year period of variability. The surface temperature is 2800K.

Ross 248 was 3 lightyears from the Solar System 36,000 years ago, and not visible then, either. Voyager 1 will pass 1.76 lightyears from it in a little over 40,000 years. No planets have been detected, but life there would be unlikely in any case.

Groombridge 34 (see #27) is the closest neighbor at a remarkably close 1.8 lightyears, followed by Krueger 60 at 4.5 lightyears. From 1.8 lightyears Ross 248 would actually be dimly visible. There are thirteen stars within ten lightyears.

This star is located at Right Ascension 23 hours 41 minutes 55 seconds, Declination +44 degrees 10' 44".

15. *Epsilon Eridani* did not need to be discovered because it is visible to the unaided eye at night, with an apparent magnitude of 3.73 and an absolute magnitude of 6.19. Its designation comes from Johann Bayer's (1572-1625) cartographic atlas, *Uranometria,* published in 1603, five years before the invention of telescopes. There is a legend that Bayer gave out the Greek letters in order of the star's brightness within each constellation, but there are many exceptions to this. Certainly Epsilon Eridani does not fit this rule, as it is the tenth brightest star in Eridanus, not the fifth as the name would suggest.

Epsilon Eridani is 10.52 lightyears away, although 105,000 years ago it was at its nearest to the Solar System at 7 lightyears. Its distance was first measured in the early 1880s by William Elkin (1855-1933). It is a member of the Ursa Major star association, believed to be the dispersed remnant of a star cluster about 500 million years old. At such a young age, Epsilon Eridani would not be expected to have yet developed advanced life forms on any planets, although Star Trek has the very advanced planet of Vulcan around this star. It has 82% of the Sun's mass, its diameter is 73.5% that of the Sun (636,000 miles), and a luminosity 34% that of the Sun. The surface temperature is 5080K. It rotates much faster than the Sun, about once every 11.2 days. As a sign of its youth, the stellar wind is thirty times the strength of the Sun's, and infrared studies indicate it has two hundred times as much dust, meteors, asteroids, and comets going around it as the Sun does. This suggests catastrophic impacts on a par with the Chicxulub impact that wiped out the dinosaurs on a fairly frequent basis, possibly as many as two or three per million years. It is classified as a K2V star.

Possibly due to its membership in a grouping of stars, it has some close approaches to other stars. 31,500 years ago it came within less than a lightyear of Luyten 726-8 (see #10 & 11 above), and 12,500 years ago to 3 lightyears of Kapteyn's Star (see #41 below). The approach to Luyten 726-8 was probably close enough to cause disturbances in the Kuiper Belt, providing an explanation for the presumed frequent impacts. There are 18 stars within ten lightyears.

In 1985 the IRAS (InfraRed Astronomy Satellite) discovered a ring of dust around Epsilon Eridani, extending from 35 to 100 AU outward. This was followed by the probable discovery of two asteroid belts and possible evidence of at least three planets. The asteroid belts are at 3 AU (280 million miles) and 20 AU (1.8 billion miles) from the star. One planet is 3.4 AU (316 million miles) from the star, taking 7 years to orbit. A possible planet is 40 AU out, taking 280 years to orbit. A proposed third planet may exist near the inner edge of the outer asteroid belt. All these planets would be large, in the range of Jupiter size. To have a climate Earth based life forms would find acceptable, a planet would have to be from 0.5 AU (47 million miles) to 1 AU (93 million miles) from the star. We have no evidence yet of any planets in that region.

Epsilon Eridani is located at Right Ascension 03 hours 32 minutes 56 seconds, Declination -09 degrees 27' 30".

16. *Lacaille 9352* was discovered by Nicholas Louis de La Caille (1713-1762) while studying the Southern Hemisphere's skies from Capetown, South Africa from 1750 to 1754. Abbé de La Caille's work was published posthumously in 1763.

This star, yet another red dwarf, has an apparent magnitude of 7.34, and an absolute magnitude of 9.75. This makes it the second brightest red dwarf in our sky. Its high proper motion was first detected by Benjamin A. Gould, who was the first American to earn a Ph.D. in astronomy (from Heidelberg University). It is 10.74 lightyears from the Solar System. From our point of view it is located in the constellation of Piscis Austrinus, the southern fish. 2700 years ago it was nearest us, but only a fraction of a lightyear closer than now. It has a mass 50% that of the Sun, a diameter 46% that of the Sun (397,000 miles), and a luminosity only 1.1% that of the Sun. The surface temperature is 3626K. No planets have been found there, but a habitable zone would be from 0.18 AU (16.7 million miles) to 0.36 AU. It is classified as M1Ve.

The closest star is the triple star system of EZ Aquarii (see #18 through 20) at 4.1 lightyears. There are sixteen stars within 10 lightyears.

Lacaille 9352 is located at Right Ascension 23 hours 05 minutes 52 seconds, Declination -35 degrees 51' 11".

17. Frank Ross returns with *Ross 128*, an M4V red dwarf he found in 1926. As a variable star, its apparent magitude goes from 11.56 to as bright as 11.06 during flares. The absolute magnitude is from 13.94 to 13.44. It is 10.92 lightyears away, but in 71,000 years it will be just 6.3 lightyears (and still invisible without optical aid, unless humans by then have considerably better eyesight). Its mass is 16% that of the Sun, the diameter 21% of the Sun (181,000 miles), and the luminosity a feeble 0.03% that of the Sun. The surface temperature is 3180K. It is in the constellation of Virgo.

Ross 128 is believed to be at least 6 billion years old. With convection ruling its interior, the core should remain well supplied with hydrogen for a very extended time, at least into the hundreds of billions of years. However, the habitable zone lies within 0.057 to 0.11 AU (about 6 million to 10 million miles). This is so close that the flares and accompanying X-Rays would make any planet in that region deadly.

Ross 128 is located at Right Ascension 11 hours 47 minutes 44 seconds, Declination +00 degrees 48' 16".

18 through 20. *EZ Aquarii* is a triple star system, also known by a couple other names, *Luyten 789-6* and *Gliese 866*. The overall apparent magnitude is 12.87. They are 11.27 lightyears from us in the constellation of Aquarius. All three are red dwarfs. Star A is classified as M5Ve, and its individual apparent magnitude is 13.33, with an absolute magnitude of 15.64. Its mass is 11% that of the Sun, and its diameter 15% that of the Sun (about 195,000 miles). Its luminosity is less than 0.2% the Sun's.

Going around Star A is Star C, dimmer at an apparent magnitude of 14.03 and an absolute magnitude of 16.34. This star is less than 720,000 miles from Star A, and orbits it in just 3.8 days in a nearly circular orbit. The mass is about 10% that of the Sun and the diameter 13% of the Sun (112,000 miles). It is classified as M6V.

Much further out, Star B orbits A and C in 2.25 years at an average distance of 113.5 million miles. This star has an apparent magnitude of 13.27 and an absolute magnitude of 15.58. This orbit is fairly elliptical, so it does get as close as 76 million miles to the inner pair. Its mass and size are similar to Star C. Star A would be a brilliant -16.13 magnitude from B.

Stars A and C are so close that they interfere with one another's zones where a habitable planet could exist. Possibly a planet could exist that orbits around both. Star B does have a viable life zone, but as with many other red dwarfs, there is the problem of occasional flares.

The closest star to this trio is Lacaille 9352 (see #16), at a distance of 4.1 lightyears, followed by Lacaille 8760 at 4.2 ly. There are fifteen stars within ten lightyears.

EZ Aquarii is located at Right Ascension 22 hours 38 minutes 33 seconds, Declination -15 degrees 18' 07".

21 & 22. *Procyon* is a pair of stars, located in the constellation of Canis Minor. They are 11.4 lightyears away, and getting closer. In 25,100 years they will be closest, at 10.3 lightyears. Star A is classified as F5, but is just about to leave the Main Sequence (V) to become a giant (IV), as its core seems to have used up its supply of hydrogen and begun to fuse helium. This star has a mass 1.4 times that of the Sun, a diameter 2.05 that of the Sun (1,760,000 miles), and a luminosity 7.73 times the Sun. The apparent magnitude is 0.38, the absolute 2.66.

Going around star A is a white dwarf, star B, in an elliptical orbit from 9 AU to 21 AU (837 million to 1.95 billion miles) away. This gives it an orbital period of 40.82 years. B has an apparent magnitude of 10.7, and an absolute of 13.04, making it very difficult to observe so close to its much brighter companion. This star was first predicted by Bessel in 1844 based on motions of the visible star A. In 1862 Georg Arthur von Auwers (1838-1915) earned his PhD by working out the detailed orbit before anyone had ever seen star B. Finally in 1896 John Schaeberle (1853-1924) saw the white dwarf. Its mass is 0.6 that of the Sun, its diameter is 11,600 miles, and luminosity is 0.055 of the Sun's.

Star B was probably originally a class B or A star which went through a giant phase and has collapsed down to a white dwarf. This would have effectively sterilized any planets around either star, assuming planets exist around either. So far none have been found. From a planet around star A the white dwarf would range from magnitude -11.41 to -13.75, while a planet near the white dwarf would see A ranging from -22.17 to -24.01.

Luyten's Star (see #37) is a mere 1.2 lightyears away, which may be close enough for mutual interference in their respective Oort Clouds. It would be dimly visible. Next in distance from Procyon is Ross 614 (see #45) at 4.6 lightyears. There are sixteen stars within ten lightyears of Procyon.

Procyon is located at Right Ascension 07 hours 39 minutes 18 seconds, Declination +05 degrees 13' 30".

23 & 24. *61 Cygni* is a pair of stars 11.4 lightyears from the Solar System in the constellation of Cygnus. They are also very rarely called *Bessel's Star*. They will be nearest the Solar System 18,000 years from now when they will be 9 lightyears away.

Star A is class K5V, with an apparent magnitude of 5.21 and an absolute magnitude of 7.49. Its mass is 70% that of the Sun, its diameter two thirds of the Sun (575,000 miles), and its luminosity is 21.5% of the Sun. The surface temperature is 4526K, and its rotation is from 27 to 45 days, depending on latitude. This star has a sunspot cycle that takes about 7.5 years.

Star B is in an elliptical orbit around A, running from 44 AU (4.1 billion miles) to 124 AU (11.5 billion miles) from its companion, taking about 678 years to complete an orbit. This star is classified as K7V, with an apparent magnitude of 6.03, an absolute magnitude of 8.31, and a mass 63% of the Sun. The diameter is 59.5% of the Sun (514,000 miles), and the luminosity is 15% of the Sun. The surface temperature is 4077K, and it rotates in 32 to 47 days, depending on latitude.

61 Cygni was the first star to have its distance measured, in 1838. It attracted special attention in the early 1950s when Swarthmore astronomer Peter van de Kamp reported finding a Jupiter sized planet around star A. However, this was eventually shown to be an error. There continue to be suspicions of one or more planets in the system, but nothing has been proven. The two stars are far enough apart so that there is plenty of room for both to have a full complement of planets, although any having an Earth-like climate would have to be within 50 to 70 million miles of their respective stars. A planet around star A would see B at magnitude -14.67 when the two stars are nearest one another (technically called periastron), and about -13.83, slightly brighter than our Full Moon, when furthest, or apastron. A would be not quite a magnitude brighter seen from a planet around B.

The closest neighbors are Krueger 60 (see #43) and V1581 Cygni (see #58), both at about 5.1 light-years. There are fifteen stars within ten lightyears.

61 Cygni is located at Right Ascension 21 hours 06 minutes 54 seconds, Declination +38 degrees 44' 44".

25 & 26. *Struve 2398* is a pair of red dwarfs in the constellation of Draco. They are 11.52 lightyears from us. Star A is classified as M3V. It has an apparent magnitude of 8.90, and an absolute magnitude of 11.16. Its mass is 36% that of the Sun, its diameter 25% of the Sun's (216,000 miles), and luminosity is 3.9% of the Sun. The surface temperature is 3500K.

At a distance of 26.26 to 85.49 AU (an average of 5.2 billion miles) Star B takes 453 years to complete an orbit around star A. It is classified as M3.5V. It has an apparent magnitude of 9.69, and an absolute magnitude of 11.95. Its mass is 30% of the Sun, with a diameter 20% of the Sun (173,000 miles), and a luminosity 2.1% of the Sun. The surface temperature is 3000 degrees.

No planets have been detected around either star. However, a planet around B would see star A as a point source at magnitude -9.97 at apastron and -12.53 (about the same as the Full Moon for us) at periastron. B would be nearly another magnitude fainter from planets around A. This is plenty bright enough to attract attention to its movement of nearly a degree a year across the sky, especially since each would be visible in the daytime.

The closest stars are BD+68 946 (see #53) at 4.2 lightyears and V1581 Cygni (see #57) at 5.8 lightyears. There are nine stars within ten lightyears.

Struve 2398 is located at Right Ascension 18 hours 42 minutes 47 seconds, Declination +59 degrees 37' 49".

27 & 28. *Groombridge 34* gets its name from a catalog prepared by Stephen Groombridge (1755-1832). The catalog was published two years after his death by the then British Astronomer Royal, George Airy (1801-1892). This is yet another pair of red dwarfs, 11.64 lightyears from us in the constellation of Andromeda.

Star A is classed as M2V, with an apparent magnitude of 8.09, and an absolute magnitude of 10.33. Its mass is 40.4% of the Sun's mass, its diameter is 37.9% of the Sun (325,000 miles), and luminosity is 6.4% of the Sun. The surface temperature is 3730K. As a flare star it is known as *GX Andromedae*.

At a distance of 146.8 AU (about 14 billion miles) Star B is classed as M3.5V. B takes about 2600 years to orbit star A. It has an apparent magnitude of 8.09, and an absolute magnitude of 10.33. Its mass is 16.3% of the Sun, diameter is 19% of the Sun (165,000 miles), and luminosity is 4.2% of the Sun. The surface temperature is 2000K. It is also known as *GQ Andromedae*.

No planets have been detected around either star, but from a planet around A, star B would appear with a magnitude of -7.44, plenty bright enough to see in daylight. The closest neighboring star is Ross 248, a mere 1.8 lightyears away. Krueger 60 is 4.9 lightyears away. There are thirteen stars within ten lightyears.

Groombridge 34 is located at Right Ascension 00 hours 18 minutes 23 seconds, Declination +44 degrees 01' 23".

29, 30, 31. *Epsilon Indi* brings us a pair of brown dwarfs associated with a star. The system is 11.83 lightyears away in the southern constellation of Indus. The star, Epsilon Indi A, is classified as K5V, and has an apparent magnitude of 4.69, with an absolute magnitude of 6.89. The mass is 76.2% that of the Sun, with a diameter 73.2% of the Sun (632,000 miles), and luminosity 17% of the Sun. The surface temperature is 4630K. The star rotates a bit faster than the Sun, and is believed to be about 1.3 billion years old.

1500 AU away (about 140 billion miles) there are two brown dwarfs, called Ba and Bb. They were discovered in 2003. Ba has a mass about 47 times that of Jupiter, and a temperature of 1280K. The diameter is about 76,000 miles. It is classed as T1V, T being one of the new categories (along with L and Y) invented for brown dwarfs.

Just 2.1 AU (195 million miles) from Ba is brown dwarf Bb. It masses just 28 times that of Jupiter, and has a surface temperature of 850K, barely able to glow at all. The diameter is about 79,000 miles. It is classed as T6V.

There have been repeated suggestions, all unconfirmed, that a planet somewhat more massive than Jupiter may orbit star A at a distance around 6.5 AU and a period of 20 years. This leaves plenty of room for habitable planets falling within the region around A where Earth life would be comfortable. From such a planet the brown dwarfs would be too faint to see without a telescope.

It is of course possible for brown dwarfs to have accompanying planets. To have an acceptable temperature for Earth life, such planets would have to be extremely close to the dwarfs, and life would have to adjust to a dim and very reddish light level. The star would be visible in the daytime for such planets, with a magnitude of -8.77.

The closest neighbors are Lacaille 8760 and Lacaille 9352, each about 4.7 lightyears away. There are seventeen stars within ten lightyears.

Epsilon Indi is at Right Ascension 22 hours 03 minutes 22 seconds, Declination -56 degrees 47' 10".

32. *DX Cancri* is yet another red dwarf, about 11.85 lightyears from us in the constellation of Cancer. The difference between this distance and that of Epsilon Indi is less than the probable error of measurement, so it could actually be slightly closer than Epsilon Indi. However, for now we'll go with the best available recent measurement. (These remarks apply to many of the following stars whose distances are sufficiently similar that the possible measurement error could cause them to swap places in this list.) This star is classed as M6.5Ve, and is a flare star like most red dwarfs. The normal apparent magnitude is 14.78, and absolute magnitude is 16.98, but during flares can brighten by as much as 1.7 magnitudes. The mass is 4.08% of the Sun's mass, the diameter is eleven percent of the Sun's (95,000 miles), and luminosity is 0.13% of the Sun. It has a surface temperature of 2840K. A planet would have to be about 3.4 million miles from the star for water to be liquid on the planet's surface. At this distance the planet's "year" would take just nine hours 45 minutes to orbit the star, and the star would appear more than 31 times the size the Sun does in our sky. And those flares would singe anything trying to live on the planet.

This star is five lightyears from its closest neighbor, Procyon (see #21). There are seventeen stars within ten lightyears.

DX Cancri is located at Right Ascension 08 hours 29 minutes 50 seconds, Declination +26 degrees 46' 37".

33. *Tau Ceti* is classed as G8Vp, so it is the most Sun-like star we've seen since Alpha Centauri A (#3). It is in the constellation of Cetus, 11.89 lightyears away. The apparent magnitude of 3.49 means it is visible in Earth's night sky except in the most light polluted cities. The absolute magnitude is 5.68. Tau Ceti has a mass 78.3% that of the Sun, a diameter 79.3% of the Sun (685,000 miles), a luminosity 52% of the Sun, and a surface temperature of 5344K. It is believed to be somewhat older than the Sun, at 6 to 10 billion years. Despite this, in 2004 it was discovered that Tau Ceti has ten times as much dust, meteors, asteroids and cometary material around it as the Sun does. The region with this junk runs from ten to 55 AU from the star, with the bulk between 35 AU (3.25 billion miles) and 50 AU (4.65 billion miles). The inner edge (10 AU) would take about 33 years to orbit the star, while the outer edge (55 AU) would take around 428 years.

Tau Ceti has a lower percentage of metals in its make up than does the Sun. This is taken by some astronomers to imply a lower probability of planets, and it is true that none have been found. On the other hand, the material found in 2004 certainly suggests Tau Ceti has enough to make planets. That the star is older than the Sun and yet no Tau Cetian monsters have invaded the Solar System for fun and profit may mean there are no (habitable) planets, that the planets are hit too often by asteroids or comets for intelligent life to get started on building a civilization, or maybe we don't taste good enough and just aren't worth the effort. The core of the habitable zone is 0.74 AU (68.8 million miles) from Tau Ceti, where a "year" would be less than 240 days.

The closest neighboring stars are YZ Ceti (see #35) at 1.6 lightyears and Luyten 726-8 at 3.2 lightyears. There are seventeen stars within ten lightyears.

Tau Ceti is at Right Ascension 01 hours 44 minutes 04 seconds, Declination -15 degrees 56' 15".

34. *GJ 1061* is another of the ever popular red dwarfs, also known as *LHS 1565*. This one is located in the southern constellation of Horologium, a little known constellation invented by de La Caille when he was filling in the southern sky. The star is classified as M5.5V, with an apparent magnitude of 13.09, and an absolute magnitude of 15.26. It is 11.99 lightyears from us. The mass is 11.3% of the Sun, and the luminosity is 0.6% of the Sun. Water would be a liquid on a presumptive planet between 0.022 AU (two million miles) and 0.064 AU (5.9 million miles) from the star.

It is only 3.7 lightyears from Kapteyn's Star (see below #41), but both are so feeble that neither would be visible from the other. Struve 2398 (see #25) is 5.8 lightyears away.

GJ 1061 is located at Right Ascension 03 hours 36 minutes 00 seconds, Declination -44 degrees 30' 45".

35. *YZ Ceti* is one of those ubiquitous red dwarf flare stars. It is classed as M4.5V, with an apparent magnitude of 12.02 and an absolute magnitude of 14.17. It is 12.13 lightyears from the Solar System in the constellation of Cetus. Its mass is 8.5% of the Sun, and diameter is 20% of the Sun (173,000 miles) with a luminosity only 0.002% of the Sun. No planets have been found. It is only 1.6 lightyears from Tau Ceti (#33 on this list).

YZ Ceti is located at Right Ascension 01 hours 12 minutes 31 seconds, Declination -16 degrees 59' 56".

36. I am listing *UGPS 0722-05* at this point, although its distance is currently only roughly determined as around 10 to 13 lightyears. This is a recently discovered free floating brown dwarf in the constellation of Monoceros. It is so cool at 500K that it is tentatively classed as Y0V, astronomers having even invented the new class Y for it. The diameter is about 86,000 miles and its mass is only twenty times that of Jupiter. Luminosity is 0.000026% of the Sun, meaning that except for an occasional hot bubble rising to the surface, it does not produce enough light to see it even if you were in a close orbit around it. No planet could exist close enough to it to receive enough heat to liquify water, as such a planet would fall within the Roche limit where it would be pulled apart by the brown dwarf's gravity.

Luyten's Star and Procyon are probably the closest neighbors, and as many as 18 stars may be within ten lightyears.

UGPS 0722-05 is located at Right Ascension 07 hours 22 minutes 28 seconds, Declination -05 degrees 40' 31".

37. *Luyten's Star* is named for Dutch astronomer Willem Luyten. It is classified as M3.5V in the constellation of Canis Minor, 12.37 lightyears from us. Luyten's Star has an apparent magnitude of 9.86, and an absolute magnitude of 11.97. The mass is 25.7% of the Sun, with a diameter 30% of the Sun (260,000 miles), and a surface temperature of 3150K. The luminosity is 0.145% of the Sun's. 13,000 years ago it was nearest the Sun, just under 12 lightyears.

The region where water would be liquid for planets here ranges from 0.102 AU (9.5 million miles) to 0.198 AU (18.4 million miles). This would give a "year" of under two months, and the star would be about ten times the size the Sun has in our sky. No planets have been found.

Procyon is only 1.2 lightyears away, and so bright it would be visible in the daytime from any planet. Next closest is Ross 614 at 3.9 lightyears. There are seventeen stars within ten lightyears.

Luyten's Star is located at Right Ascension 07 hours 27 minutes 27 seconds, Declination +05 degrees 13' 33".

38. *Teegarden's Star* was discovered in an unusual manner. Bonnard J. Teegarden (1941-) led a group of astronomers who used old photographs originally made to search for Near Earth Asteroids, and examined them for any stars that might show evidence of rapid proper motion, a clue to possible proximity to the Solar System. In 2003 they found this M6.5V red dwarf in the constellation of Aries at a distance of 12.58 lightyears. It has an apparent magnitude of 17.39 and an absolute magnitude of 19.41. Its mass is 13% of the Sun's mass, the diameter is 7.5% of the Sun's (about 65,000 miles), luminosity is 0.09% of the Sun, and the surface temperature is 2800K. It also has a catalog designation of *SO 02353+1652*.

As usual with this type of star, there are flares. No planets have been detected.

The closest star, L1159-16, is 4.0 lightyears away, followed by Epsilon Eridani at 5.8 ly. There are 15 stars within ten lightyears.

Teegarden's Star is located at Right Ascension 02 hours 53 minutes 01 seconds, Declination +16 degrees 52' 53".

39 & 40. *SCR 1845-6357* is a red dwarf and brown dwarf pair in the southern constellation of Pavo, 12.59 lightyears away. They were discovered in 2006. The star is classified M8.5V (not much better than a brown dwarf itself), with an apparent magnitude of 17.39 and an absolute magnitude of 19.41. The mass is 7% of the Sun's mass, the diameter is 9.6% of the Sun's (about 83,000 miles), the luminosity is 0.04% of the Sun, and surface temperature is 2650K.

The brown dwarf is classed as T6V, with a mass 40 times Jupiter, and a surface temperature of 950K. It is 4.1 AU (380 million miles) from the star. The system is believed to be about 2.4 billion years old.

No planets have been detected. The two dwarfs are far enough apart that either could have planets in stable orbits.

SCR 1845-6357 is located at Right Ascension 18 hours 45 minutes 05 seconds, Declination -63 degrees 57' 50".

41. *Kapteyn's Star* is named for the eminent Dutch astronomer Jacobus Kapteyn (1851-1922), who also has a lunar crater and an asteroid (818 Kapteynia) named for him. In 1898 Kapteyn (the name is pronounced kap tayn) first noted the star as having an extremely high proper motion on photographs taken at an observatory in South Africa. The star is classed as M0V, and is located in the southern constellation of Pictor. The apparent magnitude is 8.84, and absolute magnitude is 10.87. It is 12.78 lightyears from the Solar System, but 10,800 years ago was only 7 lightyears away. Its mass is 28.1% of the Sun's mass, the diameter is 29.1% of the Sun's (about 250,000 miles), and the luminosity is 1.2% of the Sun. The surface temperature is 3840K.

This star has an unusual orbit, travelling retrograde, or backwards from the direction the Sun and most other stars move in their orbits around the center of our galaxy. It is about ten billion years old, and is believed to have been captured by our galaxy from the globular cluster Omega Centauri which itself is in orbit around the galaxy.

The zone where water would be liquid extends from 0.107 AU (9.4 million miles) to 0.209 AU (19.4 million miles) from the star. Given its age, it must have formed before most of the higher numbered elements had been created, so any planets would be impoverished so far as metals are concerned. No planets have been detected.

The nearest neighbor is LHS 1565 (see #34) at 3.7 lightyears. There are nine stars within ten lightyears.

Kapteyn's Star is located at Right Ascension 05 hours 11 minutes 41 seconds, Declination -45 degrees 01' 06".

42. *Lacaille 8760* is an M0V star 12.87 lightyears from us in the southern constellation of Microscopium. Its apparent magnitude is 6.67, making it the brightest red dwarf in our sky. It is just barely possible under ideal conditions for a person with superb eyesight to see this star without any optical aids. The absolute magnitude is 8.69. It was nearest the Solar System 20,000 years ago, at 12.1 lightyears. The mass is 60% of the Sun's mass, the diameter is 51% of the Sun (440,000 miles), the luminosity 0.8% of the Sun, and surface temperature is 3340K. It rotates slower than the Sun, about once in 40 days. It is believed to be about 5.6 billion years old. The orbit is rather highly tilted with respect to the plane of the galaxy, reaching as much as 1500 lightyears off the plane. The orbit is also rather eccentric, going from 21,000 to 34,000 lightyears from the center of the galaxy.

This star first attracted attention in 1875, when the German astronomer Carl Wilhelm Moesta (1825-1884) discovered its high proper motion while checking de La Caille's work. As a variable star it is known as *AX Microscopii*.

The habitable zone is from 0.272 AU (25.3 million miles) to 0.532 AU (49.5 million miles). It does not have devastating flares, so life might be possible. but no planets have yet been detected.

Epsilon Indi is 4.1 lightyears away, followed by GJ 832 at 4.2 lightyears. There are twelve stars within ten lightyears.

Lacaille 8760 is located at Right Ascension 21 hours 17 minutes 15 seconds, Declination -38 degrees 52' 03".

43 & 44. *Krueger 60* is a pair of red dwarfs 13.15 lightyears away in the constellation of Cepheus. In 88,600 years Krueger 60 will be nearest the Sun, just 6.4 lightyears, still too faint to see. Star A is classed as M3V, with an apparent magnitude of 9.79 and an absolute magnitude of 11.76. The mass is 27.1% of the Sun's mass, with a diameter 35% that of the Sun (302,000 miles), and a luminosity 1% of the Sun's. The surface temperature is 3180K.

Star B, also known as *DO Cephei,* is in a 44.67 year orbit that ranges from 9.5 AU (883 million miles) to 13.5 AU (1.26 billion miles) from star A. B is classified as M4V, with a mass 17.6% of the Sun's mass, a diameter 24% of the Sun (198,000 miles), and a luminosity 0.34% of the Sun. The surface temperature is 2890K. The apparent magnitude is 11.40, and the absolute magnitude is 13.46. B has flares that brighten it by 0.9 magnitude, probably eliminating the possibility of life on any planets, although none have been found around either star. B would be a point in the sky a bit brighter than the Full Moon for a planet going around A.

Krueger 60 is located at Right Ascension 22 hours 28 minutes 00 seconds, Declination +57 degrees 41' 45".

45. *DEN 1048-3956* is a dim red dwarf at M8.5V. It is 13.17 lightyears from us in the southern constellation of Antlia. The apparent magnitude is 17.39, and the absolute magnitude is 19.37. The mass is 8% of the Sun's mass, the diameter is under ten percent of the Sun's size, and the luminosity is 0.035% of the Sun. It has flares, and no known planets.

DEN is short for Deep Near Infrared Survey, so it is sometimes written DENIS.

DEN 1048-3956 is located at Right Ascension 10 hours 48 minutes 15 seconds, Declination -39 degrees 56' 06".

46 & 47. *Ross 614* is a pair of red dwarfs in the constellation of Monoceros, 13.47 lightyears from us. Frank Ross discovered this as a single star in 1927, and in 1936 Dirk Reuyl (1906-1972) found it was a binary. Star A is classified as M4.5V, with an apparent magnitude of 11.76 and an absolute magnitude of 13.09. The mass is 22.28% of the Sun's mass, with a diameter 21.9% of the Sun (189,000 miles), and a luminosity 0.0029 of the Sun.

Taking 11.6 years to complete an orbit, Star B is classified as M7V, with an apparent magnitude of 14.6 and an absolute magnitude of 16.7. The mass is 11.07% of the Sun, diameter is 13% of the Sun (112,000 miles), and luminosity is 0.000215 of the Sun. The stars are separated in a range from 2.387 AU to 5.313 AU (221 million to 494 million miles).

From a planet around star A, B would appear from magnitude -11.43 to -13.16, while A would be 2.84 magnitudes brighter from B's planets. The life zone for star A is only 0.01714 AU (1.6 million miles) from the star, where a planet would take only 2.19 days to complete an orbit. At this distance the star would cover 22.4 degrees in the sky (our Sun covers only half a degree from Earth). B's life zone is at 0.00463 AU (430,000 miles, not quite twice the distance between Earth and Moon). An orbit at that distance would take only 9.775 hours.

Flares have earned Ross 614 the alternative designation as a variable star of *V577 Monocerotis*. No planets have been detected.

Ross 614 is located at Right Ascension 06 hours 29 minutes 23 seconds, Declination -02 degrees 48' 50".

48. *Wolf 1061* is 13.82 lightyears away in the constellation of Ophiuchus. It is classified as M3V. The apparent magnitude is 10.07, and absolute magnitude is 11.93. The mass is 26% of the Sun, and the diameter about a quarter that of the Sun. No planets have been detected.

Wolf 1061 is located at Right Ascension 16 hours 30 minutes 18 seconds, Declination -12 degrees 39' 45".

49. *Gliese 1* is an M2V star in the constellation of Sculptor. It is 14.23 lightyears from us. The apparent magnitude is 8.55, and absolute magnitude is 10.35. It has 45% of the Sun's mass, 46% of its diameter (397,000 miles), and a surface temperature of 3380K. There are frequent flares and no known planets.

Gliese 1 is located at Right Ascension 00 hours 05 minutes 24 seconds, Declination -37 degrees 21' 27".

50 & 51. *Wolf 424* is a pair of red dwarfs in the constellation of Virgo. They are 14.31 lightyears from the Solar System. The system is also known as *LHS 333* and *GJ 473*.

Star A is classified as M5.5V, with an apparent magnitude of 13.18, and an absolute magnitude of 14.97. Its mass is 14% of the Sun's mass, with a diameter 17% of the Sun's (147,000 miles). The life zone around star A runs from 0.022 AU to 0.054 AU (2 million to 5 million miles).

Star B is classified as M7Ve, with an apparent magnitude (ignoring the effects of frequent flares and sunspots) of 13.17, and an absolute magnitude of 14.96. The mass is 13% of the Sun's mass, with a diameter 14% of the Sun (121,000 miles). The luminosity is 0.008 of the Sun. B is 2.36 to 4.19 AU (220 to 390 million miles) from star A, with an orbit taking 16.2 years. Star B is a UV Ceti type variable, with the designation *FL Virginis*.

From a planet around star A, B would range from -13.33 to -14.48, probably bright enough to read by, although cautious mommies will warn "you are hurting your eyesight reading by B light!" From a planet around B, A would be slightly brighter. The life zone around A is just 0.00991 AU (920,000 miles), where an orbit would take 1.34 days and A would cover 8.9 degrees in the sky.

Ross 128 is the closest neighbor, 4.0 lightyears away. There are eleven stars within ten lightyears.

Wolf 424 is located at Right Ascension 12 hours 33 minutes 17 seconds, Declination +09 degrees 01' 16".

52. *Van Maanen's Star* is a white dwarf in the constellation of Pisces, discovered by Adriaan van Maanen in 1917. The name was introduced by Willem Luyten when he did his initial studies of white dwarfs. This was the third white dwarf to be discovered, and the first that did not have a stellar companion. Its apparent magnitude is 12.36, and absolute magnitude is 14.21. Its distance is about 14.32 lightyears. It has 70% of the Sun's mass, but a diameter of only 8700 miles.

Several studies, each inconsistent with the others, have claimed to detect a planet near this star, but none have been confirmed as valid. This star has more stellar neighbors within ten lightyears than any other star on this list, 24 known at last count. Yet there is no star closer to it than 4.3 lightyears.

Van Maanen's Star is located at Right Ascension 00 hours 49 minutes 10 seconds, Declination +05 degrees 23' 19".

53. *HIP 15689* is in the constellation of Eridanus, with an absolute magnitude of 13.94 and an apparent magnitude of 12.16. The luminosity is only 0.023% that of the Sun. The closest neighboring star is Epsilon Eridani at 4.1 lightyears. There have been suggestions that this may be a binary star, or that it has a brown dwarf companion, but this is unconfirmed. It is classified as M6.5V.

HIP 15689 is located at Right Ascension 03 hours 22 minutes, Declination -13 degrees 16' 40".

54. *TZ Arietis* is an M4.4V flare star in Aries. Its apparent magnitude is 12.27, and an absolute magnitude of 14.03. It is 14.51 lightyears from us. Its mass is 8.9% of the Sun's mass, the diameter is under ten percent that of the Sun, and the luminosity is 0.021% of the Sun. The surface temperature is about 3100K.

TZ Arietis is located at Right Ascension 02 hours 00 minutes, 13 seconds, Declination +13 degrees 03' 08".

55. *BD +68 degrees 946* is an M3V star in the constellation of Draco. The apparent magnitude is 9.17, the absolute magnitude is 10.89, and it is 14.79 lightyears from us. The mass is 21% of the Sun's mass, but it has a much higher percentage of elements higher than helium than the Sun does. This is generally taken to mean that the star is significantly younger than the Sun. Its diameter is 49.2% of the Sun (425,000 miles). It is a prolific producer of X-rays (also suggestive of youthfulness), discouraging any Earth type life. No planets have been found.

This star is also known as *LHS 450, HIP 86162, AOe 17415.6,* and as *GJ 687.*

BD +68 946 is located at Right Ascension 17 hours 36 minutes 26 seconds, Declination +68 degrees 20' 21".

56. *LHS 292* first appeared in a catalog prepared by Willem Luyten. It is an M6.5V star in the constellation of Sextans, 14.81 lightyears from us. Outside of its frequent flares it has an apparent magnitude of 15.6, and an absolute magnitude of 17.32. The mass is 17% of the Sun's mass, the diameter 11.7% of the Sun (101,000 miles), and the luminosity is 0.069% of the Sun. The surface temperature is 2750K.

LHS 292 is located at Right Ascension 10 hours 48 minutes 13 seconds, Declination -11 degrees 20' 14".

57. *GJ 674* is an M3V star in the constellation of Ara, about 14.81 lightyears away, discovered around 1892 by John A. Thome (1843-1908) from Cordoba Observatory in Argentina. The apparent magnitude is 9.38, and the absolute magnitude 11.09. The mass is 35% of the Sun, the diameter 42% of the Sun (363,000 miles), and the luminosity 0.32% of the Sun. The surface temperature is 3600K. Its rotational period is 34.8 days, and it is believed to be about 550 million years old. It is also known as *CD -46 degrees 11540, LHS 449,* and *HIP 85523.*

GJ 674 has a planet discovered in 2007 by Xavier Bonfils (1960-) and colleagues at the European Southern Observatory in La Silla, Chile. The planet's mass is twelve times that of Earth. We have no planet is the Solar System quite matching that mass, so it is hard to predict what it would be like. It is 0.039 AU (3.6 million miles) from the star, taking 4.7 days to orbit it. This is actually too close for liquid water, which would be found at 0.128 to 0.149 AU (11 to 14 million miles).

There are twenty stars within ten lightyears, the closest just 1.8 ly.

GL 674 is located at Right Ascension 17 hours 28 minutes 40 seconds, Declination -46 degrees 53' 43".

58. *WISE J1741 +2553* is a brown dwarf whose discovery by a team of German astronomers was reported in May 2011. Ralf-Dieter Scholz, G. Bihain, O. Schnurr, and J. Storm of the Leibniz-Institut fuer Astrophysik Potsdam used NASA's WISE spacecraft. The object is in the constellation of Hercules, about 15 lightyears away, classified as T9.5V. Its mass is only 7.5% that of the Sun. (WISE stands for Wide-field Infrared Survey Explorer, the "explorer" continuing a series that goes back to America's very first satellite, Explorer 1, launched January 31, 1958.)

WISE J1741 +2553 is located at Right Ascension 17 hours 41 minutes, Declination +25 degrees 53'.

59, 60, 61. GJ 1245 is a star system also known as *V1581 Cygni* in the constellation of Cygnus, about 14.82 lightyears away. `Star A is classified as M5.5V, with an apparent magnitude of 13.46, and an absolute magnitude of 15.17. Its mass is just 11% of the Sun's mass, and the luminosity is 0.84% of the Sun. The life zone is only 0.00915 AU (850,000 miles) from the star.

Less than 8 AU (700 million miles) from star A we have the M7V star C, with a mass 7% of the Sun's. It has an apparent magnitude of 16.75, and an absolute magnitude of 18.46.

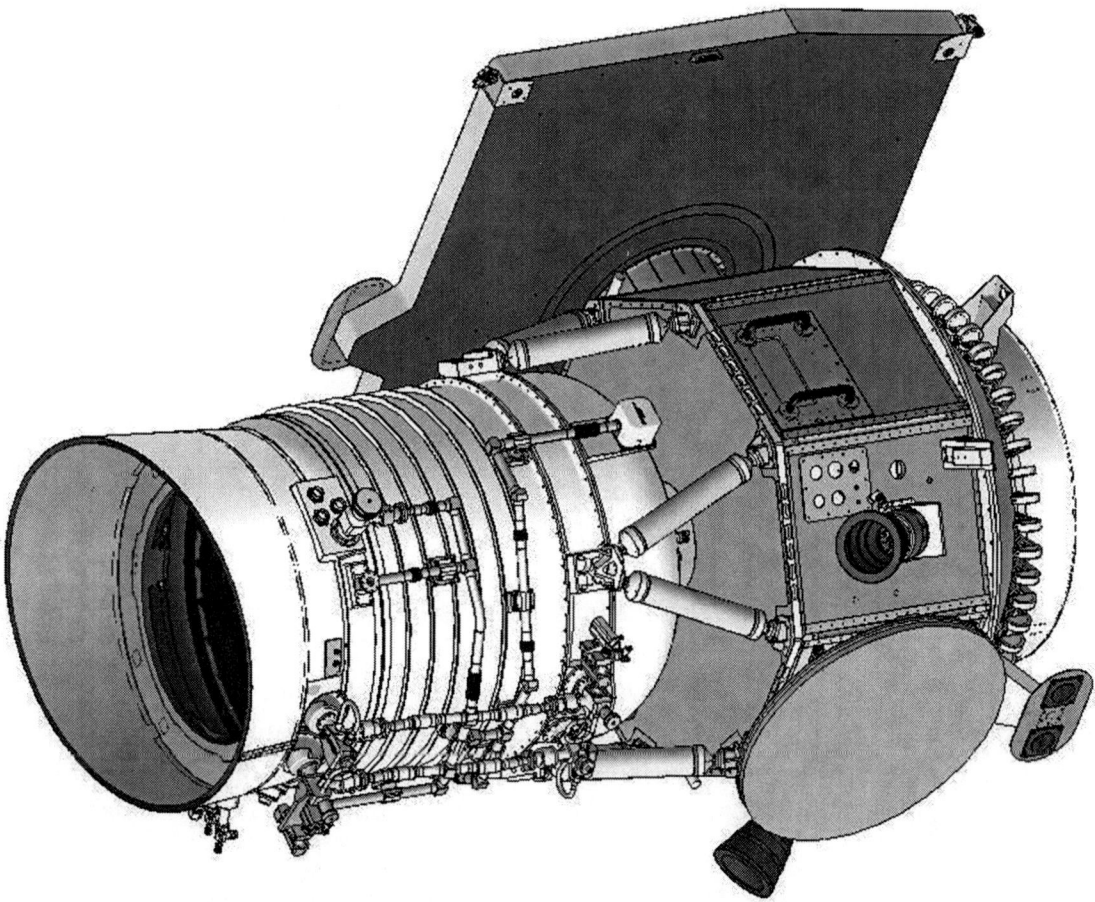

The WISE spacecraft, used by NASA to study cool objects in space.

At a distance of 33 AU (a little over 3 billion miles) we have star B, classified as M6V. It has an apparent magnitude of 14.01, and an absolute magnitude of 15.72. The luminosity is 0.048% of the Sun, and the mass is 10% of the Sun. No planets have been detected in this system. A planet around this star would see A as magnitude -8.94, and C at about the brightness we see Venus in our sky.

GJ 1245 is located at Right Ascension 19 hours 53 minutes 54 seconds, Declination +44 degrees 24' 55".

62. *GJ 440* is a white dwarf 15.06 lightyears from us in the constellation of Musca. The apparent magnitude is 11.5, and the absolute magnitude is 13.18. The mass is 75% of the Sun's mass, but the diameter is only 1% of the Sun (8600 miles, barely larger than the Earth). This gives it a luminosity only 0.05 of the Sun, despite a temperature of 8500K. It is believed to have collapsed down to its current white dwarf state 1.44 billion years ago, after having been a B star, probably somewhere in the range of B4V to B9V, with a mass 4.4 times that of the Sun. The dispersal into space of so much hot material would have sterilized, if not destroyed, any planets. None are known, but for the record the current habitable zone would lie 0.0216 AU (2 million miles) from the star. Anything that close to a main sequence B star would at the least be molten. This star is also known as *LHS 43* and *HIP 57367*.

GJ 440 is located at Right Ascension 11 hours 45 minutes 43 seconds, Declination -64 degrees 50' 29".

63. *GJ 1002* is a red dwarf classified as M5.5V at a distance of 15.31 lightyears in the constellation of Cetus. The apparent magnitude is 13.76, and the absolute magnitude is 15.40, with 11% of the Sun's mass. For a red dwarf it is remarkably stable, with few or no flares. This suggests a great age. No planets have been reported.

GJ 1002 is located at Right Ascension 00 hours 06 minutes 44 seconds, Declination -07 degrees 32' 22".

64. *Ross 780* is a red dwarf classified as M3.5V, 15.34 lightyears away in the constellation of Aquarius. The apparent magnitude is 10.17, and the absolute magnitude is 11.81. The mass is 33.4% of the Sun, the diameter is 36% of the Sun (335,000 miles), with a luminosity 0.24% of the Sun, and a surface temperature of 3350K. It has large sunspots, emits x-rays, and is believed to be about 7 billion years old.

In 1998 Geoffrey Marcy (1954-) and colleagues discovered a planet around Ross 780. Given the label b, it goes around the star in 61.1 days at a distance of 0.21 AU (19.5 million miles). It has a mass 2.28 times that of the planet Jupiter, which itself has a mass 318 times the Earth's mass, so this is no lightweight.

A second planet was found in 2001. Labelled c, it is 0.13 AU from the star (12 million miles), with a period of 30 days. It has a mass 71% of Jupiter.

The third planet was labelled d when it was discovered in 2005. It is the closest to Ross 780, at 0.02 AU (not quite 2 million miles), with a period of 1.94 days. Its mass is 6.8 times that of Jupiter.

The last planet (so far) was discovered in 2010, and got the label e. It is furthest from the star at 0.33 AU (31 million miles), with a period of 124.3 days, and is also the most massive planet at 14.6 times the mass of Jupiter. (This actually puts it fairly close to the minimum for brown dwarfs.)

Clearly none of these planets would support Earth style life, even ignoring the low heat output, flares and x-rays of the star. However, a moon of the planet nearest the star, d, would fall near the region where water could be liquid, and if the planet has a magnetic field strong enough to protect its moons from the x-rays and atomic particles released by flares, life is just barely conceivable.

EZ Aquarii (see #18) is 4.2 lightyears away, and there are fourteen stars within ten lightyears.

Ross 780 is located at Right Ascension 22 hours 53 minutes 17 seconds, Declination -14 degrees 15' 49".

65. *LHS 288* is from a catalog created by Willem Luyten. It is an M5.5V red dwarf in the constellation of Carina, 15.61 lightyears from us. The apparent magnitude is 13.90, and the absolute magnitude is 15.51. It has 11% the mass of the Sun. Its nearest neighbor is DEN 0817-6155 (see #69).

In 2006 it was discovered that this star has a planet 2.4 times the mass of Jupiter. This planet takes 6.8 years to complete an orbit.

LHS 288 is located at Right Ascension 10 hours 44 minutes 21 seconds, Declination -61 degrees 12' 36".

66 & 67. *GJ 412* is a pair of red dwarfs 15.83 lightyears from us in the constellation of Ursa Major. Star A is classified as M1V, with an apparent magnitude of 8.77, and an absolute magnitude of 10.34. Its mass is 48% of the Sun's mass, and the surface temperature is 3687K.

Star B is also known at *WX Ursae Majoris*, as an indication of its variability. It is classified as M5.5V. The mass is 10% of the Sun. In 1939 Adriaan van Maanen (1884 - 1946) discovered flares that noticeably brightened it from its normal apparent magnitude of 14.48 and absolute magnitude of 16.05. The flares include strong releases of x-rays. The surface temperature away from flares is 2700K. It is 190 AU (17.7 billion miles) from star A. At this distance any planets around A would be unaffected by the flares and x-rays, and the star itself would appear less bright than the Full Moon does to us.

Groombridge 1618 (see #68) would be a third magnitude star, and closest neighbor, for any planets in this system. There are ten stars located within ten lightyears of this pair.

GJ 412 is located at Right Ascension 11 hours 05 minutes 29 seconds, Declination +43 degrees 31' 27".

68. *Groombridge 1618* is a K7V star 15.85 lightyears from us in Ursa Major. Its apparent magnitude is 6.59, and absolute magnitude is 8.16. The mass is 64% of the Sun's mass, with a diameter 59% of the Sun's (about 550,000 miles), and a luminosity 4.6% of the Sun. The surface temperature is 3970K. It has some flares and sunspots, but is not very active. The star is believed to about one billion years old. It was one of the earliest stars to have its distance measured, in 1884. This star is also known as *Gliese 380* and *HIP 49908*.

The habitable zone for this star ranges from 0.354 AU to 0.691 AU. A planet is suspected but not confirmed at 0.41 AU from the star (38 million miles), with a period of 122 days. It would have a mass about four times that of Jupiter.

The closest neighboring stars are WX Ursae Majoris (see #67) and its companion, 3.1 lightyears away. At this distance WX would flicker in and out of visibility from the effects of its flares. There are ten stars within ten lightyears.

Groombridge 1618 is located at Right Ascension 10 hours 11 minutes 22 seconds, Declination +49 degrees 27' 15".

69. *AD Leonis* is an M4V red dwarf in the constellation of Leo, 15.94 lightyears from us. It is one of the most active of flare stars, with x-rays and pronounced sunspots, but in its quiet stage has an apparent magnitude of 9.32, and an absolute of 10.87. Its mass is 40% of the Sun's, a diameter 39% of the Sun

(337,000 miles, or 2.6 X 10^8 meters), and a luminosity 2.4% of the Sun. The surface temperature away from the flares and spots (respectively hotter and cooler) is 3450K. No planets have been detected.

AD Leonis is located at Right Ascension 10 hours 19 minutes 36 seconds, Declination +19 degrees 52' 10".

70. *DEN 0817-6155* is a brown dwarf 16 lightyears away in the constellation of Carina. It is classified as T6V. It was discovered by a team consisting of Etienne Artigau, Jacqueline Ratigan, Stuart Folkes, and others.

DEN 0817-6155 is located at Right Ascension 08 hours 17 minutes 30 seconds, Declination -61 degrees 55' 12".

71. *GJ 832* is an M1.5V star in the constellation of Grus, 16.09 lightyears from us. It has an apparent magnitude of 8.66, and an absolute magnitude of 10.20. The mass is 45% of the Sun, with a diameter 48% the size of the Sun (415,000 miles), and a luminosity 2.8% of the Sun. The surface temperature is 3620K. It is also known as *CD -49 degrees 13515.*

In 2008 Jerome Bailey and a team of astronomers working at the European Southern Observatory in Siding Spring, Australia discovered a planet with 64% the mass of Jupiter in an orbit 3.4 AU (316 million miles) from this star. It takes 3416 days (9.36 years) to complete an orbit.

There are fourteen stars within ten lightyears. The closest is Lacaille 8760 (see #42 above) at 4.2 lightyears, followed by Epsilon Indi at 4.8.

GJ 832 is located at Right Ascension 21 hours 33 minutes 34 seconds, Declination -49 degrees 00' 32".

72. *LP 944-020* is a brown dwarf, classified as T3V. It was discovered by Luyten (the LP in its name is for "Luyten Palomar", where he was working). It is in the constellation Fornax, 16.20 lightyears from us. The apparent magnitude is 18.50, and the absolute magnitude 20.02. Its mass is less than 8% of the Sun's mass. The diameter is about 86,000 miles.

Lithium and clouds of some sort were detected in its atmosphere in 2007. The presence of lithium is determinative for a classification as a brown dwarf. On December 15, 1999 a flare roughly equal in strength to a moderate solar flare was detected by the Chandra Space Telescope, showing there is some activity even in these cool objects. It rotates in less than five hours, a remarkable speed for such a large object.

The nearest star to this brown dwarf is 82 Eridani (outside the range covered in this book) at 4.4 lightyears, followed by GJ 1061 (see #34 above) at 4.9 lightyears. There are eleven stars within ten lightyears. Its proper motion suggests it might be a member of a star cluster that includes the bright star Castor.

LP 944-020 is located at Right Ascension 03 hours 39 minutes 35 seconds, Declination -35 degrees 25' 41".

73. *DEN 0255-4700* is the nearest class L brown dwarf to the Solar System, being L7.5V at a distance of 16.20 lightyears. It is in the constellation of Eridanus. The apparent magnitude is 22.92, with an absolute magnitude of 24.44. The mass is only 7% of the Sun's mass, and the surface temperature is 1700K, fairly warm for a brown dwarf. It was identified as nearby in 2006 by Edgardo Costa and Rene Mendez, members of the RECONS team working at Cerro Tololo Observatory in Chile.

DEN 0255-4700 is located at Right Ascension 02 hours 55 minutes 35 seconds, Declination -47 degrees 00' 52".

74 to 76. *Keid,* also known as *40 Eridani* (or *Omicron-2 Eridani*), is a triple star system 16.24 lightyears from us in the constellation Eridanus. The name Keid is from the Arabic word for eggshell. The group is also known as *LP 656-38.*

Star A is classified as K0.5V, with an apparent magnitude of 4.43 and an absolute magnitude of 5.94. Its mass is 89% that of the Sun. This star has a habitable zone around 0.5 AU. No planets have been detected. William Herschel first determined that Keid was a binary star in 1782. Around 1850 Otto Wilhelm von Struve (1819-1905) found the dimmest star, giving us a triple star system.

At a distance of about 50 AU from star A we have its dimmer companion, Star C. This is a red dwarf, M4V, with an apparent magnitude of 11.24 and an absolute magnitude of 12.75. Its mass is 18% of the Sun. It would be a point or dot almost as bright as the Full Moon is to us from a planet in orbit around star A. As a UV Ceti type flare star, it is also known as *DY Eridani.*

At a remote 600 AU we find a white dwarf as star B, with an apparent magnitude of 9.52 and an absolute magnitude of 11.03. Its mass is more than half that of the Sun. From a planet around either of the other two stars it would appear as a point about two magnitudes brighter than Venus is at its brightest in our sky, and therefore visible in the daytime. It probably was around B8 to A2 when it was a Main Sequence star. It is not known if it has any planets that survived its red giant phase.

The nearest neighbor is HIP 15689 (#52 above) at 4.4 lightyears. There are fifteen stars within ten lightyears.

40 Eridani is located at Right Ascension 04 hours 15 minutes 16 seconds, Declination -07 degrees 39' 10".

77. *GJ 873* This red dwarf is 16.47 lightyears from us in the constellation Lacerta. It is classified as M3.5V, and has a mass 28% that of the Sun. The diameter is 38% of the Sun's. The apparent magnitude is 10.22, and the absolute magnitude is 11.71. It rotates in just under 4.4 days. No planets have been detected. In the 1980s several astronomers suggested there is an unseen companion with a 45 year period and a mass around 2% that of the Sun. Such an object would be a brown dwarf, but remains unconfirmed.

As a variable, this star is also known as *EV Lacertae,* and is a UV Ceti type. On April 25, 2008 NASA's Swift Satellite picked up the most powerful flare ever seen on any star, thousands of times more powerful than any ever seen on the Sun. This is taken as part of the proof that this is a very young star, with an exceptionally strong magnetic field about 100 times stronger than the Sun's. Life would be fried by such powerful flares, but for the record, the Life Zone is about 0.112 to 0.218 AU from the star.

The closest star is Krueger 60 (see #43 above), at a distance of about 4.9 lightyears. There are fourteen stars within ten lightyears.

GJ 873 is located at Right Ascension 22 hours 46 minutes 50 seconds, Declination +44 degrees 20' 02".

78. *GJ 682* is a red dwarf classified as M4V in the constellation of Telescopium. Its mass is 20% that of the Sun, with an apparent magnitude of 10.95 and an absolute magnitude of 12.43. It is 16.48 lightyears away. No planets have been found.

GJ 682 is located at Right Ascension 17 hours 37 minutes 04 seconds, Declination -44 degrees 19' 09".

79 & 80. *70 Ophiuchi* is a pair of orange stars 16.64 lightyears away in the constellation of Ophiuchus. While they are visible to the naked eye as a single star, late in the Eighteenth Century Wlliam Herschel

(1738 -1822) discovered that this is a binary. Herschel is better known for his discovery of the planet Uranus.

Star A has an apparent magnitude of 4.21, with an absolute magnitude of 5.67. It is classified as K0V, with a mass 92% of the Sun, a diameter 89% of the Sun (769,000 miles), and a luminosity 43% of the Sun.

Star B has an apparent magnitude of 6.01, with an absolute magnitude of 7.47. It is classified as K5V, with a mass 70% that of the Sun, and diameter 73% of the Sun (630,000 miles). It is in an elliptical orbit ranging from 11.4 AU to 34.8 AU (1.06 billion miles to 3.25 billion miles) from star A, taking 83.38 years to complete an orbit.

There have been repeated claims of detecting a planet around one or both stars, going all the way back to 1896, when Thomas Jefferson Jackson See (1866-1962) made the first such claim. None have been substantiated. Star A would be a very bright -20.58 magnitude from a planet around star B when the two stars are closest (periastron). Star B would be -16.40 from a planet around A when they are furthest apart (apastron).

Star A's life zone is centered on 0.683 AU (63.5 million miles) taking 227.6 days. B's is at 0.2995 AU (27.8 million miles) in a 77.3 day orbit.

Ross 652 (see #100 below) and van Biesbroeck's Star (see #101 below) are each 6.1 lightyears from 70 Ophiuchi, with fifteen stars within ten lightyears.

70 Ophiuchi is located at Right Ascension 18 hours 05 minutes 27 seconds, Declination +02 degrees 30' 00".

81. *Altair* is the twelfth brightest star in our sky, with an apparent magnitude of 0.77, and an absolute magnitude of 2.22, 16.77 lightyears from us. It is classified as A7V-IV, in the constellation of Aquila. (Aquila is the Latin word for eagle, while Altair is from the Arabic for vulture.) The classification showing it is between two Roman numerals suggests that Altair is just about finished using up its hydrogen, and moving off the Main Sequence (shown as V) onto the Giant Branch (IV). Since its age is estimated at about one billion years, this demostrates how massive stars evolve more quickly, as Altair has a mass 1.71 times that of the Sun. Altair rotates in the remarkably fast time of 8.9 hours, which noticeably distorts its shape. The equatorial diameter of this star is 2.03 times that of the Sun (1.754 million miles), but the diameter through the poles is only 1.63 times the size of the Sun (1.408 million miles). This even affects the surface temperature, which is 8500K near the poles, but only 6900K along the equator. The flattening was discovered by Gerard van Belle (1968-) and colleagues working at Mount Palomar Observatory in 1999 and 2000.

No planets have been found near Altair, and there are some theoretical reasons for believing a star rotating so rapidly may not have planets.

Altair is also part of an asterism, an informal group of stars, known as the Summer Triangle. The other members are Vega (in Lyra, about 26.5 lightyears from us), and Deneb (in Cygnus, and very far away).

Ross 652 & van Biesbroeck's Star are 3.7 lightyears from Altair, followed by 70 Ophiuchi at 7.8. There are only nine stars within ten lightyears.

Altair is located at Right Ascension 19 hours 50 minutes 47 seconds, Declination +08 degrees 52' 06".

82 & 83. *GJ 1116* is a pair of red dwarfs in the constellation of Cancer, 17.30 lightyears from us.

Star A has an apparent magnitude of 14.06, with an absolute magnitude of 15.46. It is classified as M5.5V, with a mass just 11% of the Sun. It has many flares, leading to its designation as a variable star of *El Cancri*.

About 23.5 AU away (2.2 billion miles) we have Star B, an even more feeble red dwarf with an apparent magnitude of 14.92 and an absolute magnitude of 16.32. Its mass is 10% of the Sun, and it too has flares. Planets around B would see Star A as magnitude -9.25, while B would be -8.39 from planets around A. A's life zone is only 0.0075 AU or less than 700,000 miles from the star, with an orbital period of 23.14 hours. No planets have been detected around either star.

GJ 1116 is located at Right Ascension 08 hours 58 minutes 15 seconds, Declination +19 degrees 45' 47".

84. *G 099-049* is a red dwarf in the constellation of Orion. It is classified as M4V, with an apparent magnitude of 11.33 and an absolute magnitude of 12.68. This star has a mass 15% that of the Sun. It has a diameter 10.5% the Sun's (91,000 miles), and is 17.52 lightyears from us. No planets have been detected.

G 099-049 is located at Right Ascension 06 hours 00 minutes 03 seconds, Declination +02 degrees 42' 24".

85. *LHS 1723* first appeared in a catalog prepared by Willem Luyten. It is classified as M4V in the constellation of Eridanus. The apparent magnitude is 12.22, and the absolute magnitude is 13.59. It is 17.54 lightyears from us. The mass is 15% of the Sun's mass.

LHS 1723 is located at Right Ascension 05 hours 01 minutes 57 seconds, Declination -06 degrees 56' 47".

86. *2MA 0939-2448* is a brown dwarf classified as T8V, in the constellation of Puppis. The mass is only 3% of the Sun.

2MA 0939-2448 is located at Right Ascension 09 hours 39 minutes 36 seconds, Declination -24 degrees 48' 28".

87. *GJ 445* is classified M3.5V, located in the constellation of Camelopardalis. It is 17.55 lightyears from us, with a mass 23% that of the Sun, an apparent magnitude of 10.79, and an absolute magnitude of 12.15. The surface temperature is 3260K. The luminosity is just 0.8% of the Sun. The diameter is 32% of the Sun (276,000 miles). The star emits x-rays along with flares.

In about 40,000 years the star will be at its closest to the Solar System, 3.45 lightyears, and at the same time Voyager 1 will pass 1.6 lightyears from the star. It will still be below naked eye visibility from Earth.

GJ 445 is located at Right Ascension 11 hours 47 minutes 41 seconds, Declination +78 degrees 41' 28".

88. *Wolf 498* is classified as M1.5V. It is in the constellation Hercules, with an apparent magnitude of 8.46, and an absolute magnitude of 9.79, 17.83 lightyears from us. The mass is 54% of the Sun's mass, with a diameter 48% of the Sun (415,000 miles). The luminosity is 3.7% that of the Sun, and the surface temperature is 3670K.

Wolf 498 is located at Right Ascension 13 hours 45 minutes 44 seconds, Declination +14 degrees 53' 30".

89 & 90. *GJ 169* is a pair of stars in the constellation Camelopardalis. They are 17.6 lightyears away from us.

Star A is classified as M4V, with an apparent magnitude of 11.04, and an absolute magnitude of 12.32. Its mass is 20% of the Sun's mass., and its diameter is 25% the size of the Sun (216,000 miles).

Star B is a white dwarf about 9 AU (840 million miles) from its companion. It has an apparent magnitude of 12.43, and an absolute magnitude of 13.71. Its mass is 75% of the Sun's mass, but its diameter is 12,000 miles. It would be slightly brighter than the Full Moon is to us from a planet going around Star A.

GJ 169 is located at Right Ascension 04 hours 31 minutes 12 seconds, Declination +58 degrees 58' 38".

91. *GJ 251* is another red dwarf, classified as M3V in the constellation Auriga. It has an apparent magnitude of 10.02, and an absolute magnitude of 11.27. Its mass is 34% that of the Sun. It is 17.65 lightyears from us.

GJ 251 is located at Right Ascension 06 hours 54 minutes 49 seconds, Declination +33 degrees 16' 05".

92. *2MA 1835+3259* is barely warm enough to be classified as a red dwarf, at M8V, with a mass just 7% of the Sun's mass, and an apparent magnitude of 18.27. The absolute magnitude is 19.50. It is 17.75 lightyears from us in the constellation Lyra.

2MA 1835+3259 is located at Right Ascension 18 hours 35 minutes 38 seconds, Declination +32 degrees 59' 54".

93. *Wolf 1453* has a status that is more disputed than most stars. SIMBAD, a major astronomical group based in Strasbourg, France, says this star is K2V, agreeing with the Henry Draper catalog. However, the Research Consortium on Nearby Stars (RECONS), based at Georgia State University, says that it is M1.5V. They also do not agree on the apparent magnitude, which is very odd, RECONS giving 7.95, while SIMBAD says 8.80. The absolute magnitude is either 9.90 (SIMBAD) or 9.18 (RECONS).

This ambiguous star is 17.88 lightyears away in the constellation of Orion. Its mass is probably 57% of the Sun's mass, which would lean towards RECONS being correct.

Wolf 1453 is located at Right Ascension 05 hours 31 minutes 27 seconds, Declination -03 degrees 40' 38".

94. *LP 816-060* is an M1V star in Capricorn. The apparent magnitude is 11.50, and absolute magnitude is 12.72. The mass is 19% of the Sun. The apparent magnitude is 11.50, and the absolute magnitude is 12.72. The star is 17.9 lightyears from us.

LP 816-060 is located at Right Ascension 20 hours 52 minutes 33 seconds, Declination -16 degrees 58' 29".

95. *WISE J0254+0223* is a brown dwarf discovered in 2011 by Ralf-Dieter Scholtz and his team. It is classified as T9V, and is about 18 lightyears away. Methane has been detected in its atmosphere.

WISE J0254 +0223 is located in the constellation of Cetus, at Right Ascension 02 hours 54 minutes, Declination +2 degrees 23'.

96. *2MA 0415-0935* is a T8V brown dwarf in the constellation Eridanus. It has a mass just 3% that of the Sun, and is 18.7 lightyears from us.

2MA 0415-0935 is located at Right Ascension 04 hours 15 minutes 20 seconds, Declination -09 degrees 35' 07".

97. *Alsafi,* or *Sigma Draconis,* is a G9V star in the constellation of Draco. The name Afsafi means "tripods" in Arabic, referring to burners used in desert cooking. The star is 18.79 lightyears from the Solar System, with an apparent magnitude of 4.68 (dimly visible to the naked eye), and an absolute magnitude of 5.87. The mass is 89% of the Sun's mass, with a diameter 77.8% of the Sun (672,000 miles), and a luminosity 42.8% of the Sun. The surface temperature is 5350K. Alsafi rotates once in 29 days. It is believed to be about 7 to 8 billion years old, slightly older than the Sun. No planets have been discovered, but the life zone would be about 0.557 AU to 0.95 AU (52 million to 90 million miles) from the star.

The closest neighbor to Alsafi does not make this list of our neighbors, but for the record, LP 44-113 is 3.2 lightyears away from Alsafi. BD +68^946 (#54 above) is 5 lightyears away from Alsafi. There are nineteen stars within ten lightyears.

Alsafi is located at Right Ascension 19 hours 32 minutes 22 seconds, Declination +69 degrees 39' 40".

98 & 99. *GJ 229* is a red dwarf and brown dwarf pair in the constellation of Lepus. They are 18.8 lightyears from us. The red dwarf is also named *HIP 29295* and *HD 42851.*

The red dwarf is classified as M1.5V, with a mass 56% of the Sun's mass, and an apparent magnitude of 8.14, with an absolute magnitude of 9.34. The habitable zone where water could be liquid is 0.231 to 0.450 AU. The star has 1.6% the luminosity of our Sun.

The brown dwarf was discovered October 27, 1994, and is classified as T6V. It has a mass 3% of the Sun, and a diameter about the same size as Jupiter. It is 39 AU (3.63 billion miles) from the red dwarf, taking about 340 years to orbit the star. It would be too faint to see without optical aids from a planet around the red dwarf. A planet around the brown dwarf would see the star as a point source somewhat less bright than the Full Moon is on Earth.

The closest star to GJ 229 does not make the list of close stars to us, but for the record, it is BD -03 1123 at 6.7 lightyears. G 99-44 (#83 above) is 6.8 ly from this binary. There are eleven stars within ten lightyears.

GJ 229 is located at Right Ascension 19 hours 32 minutes 22 seconds, Declination -21 degrees 51' 53".

100. *Ross 47* is classified as M4V, and is in the constellation of Orion. It is 18.9 lightyears away from us. The apparent magnitude is 11.57, and the absolute magnitude is 12.74. The mass is 20% the mass of the Sun. It has many flares.

Ross 47 is located at Right Ascension 05 hours 42 minutes 09 seconds, Declination +12 degrees 29' 22".

101. *GJ 693* is an M3V star in the constellation of Pavo. It is 19.0 lightyears from us. The apparent magnitude is 10.76, with an absolute magnitude of 11.93. The mass is 26% of the Sun.

GJ 693 is located at Right Ascension 17 hours 46 minutes 34 seconds, Declination -57 degrees 19' 09".

102 & 103. *GJ 752* is a pair of red dwarfs in the constellation of Aquila. They are 19.2 lightyears from us. They are believed to be only one billion years old.

Star A is also known as *Wolf 1055.* It is classified as M3.5V, with an apparent magnitude of 9.10, and an absolute magnitude of 10.26. Its mass is 49% of the Sun's mass, with a diameter 52.4% of the Sun's (453,000 miles) and a luminosity just 0.2% of the Sun. Its surface temperature is 3240K. A planet with less than six times the mass of Jupiter is suspected at a distance of 0.333 AU (31 million miles), taking 0.744 Earth year to orbit the star.

In 1940 George van Biesbroeck (1880-1974) discovered the companion star B at a distance of 434 AU (over 40 billion miles). This star is now known, among other designations, as *van Biesbroeck's Star.* It is classified as M8Ve, and while the normal surface temperature is 2700K, it has many powerful flares. The apparent magnitude is 17.45, and the absolute magnitude is 18.61. The mass is just 7.5% of the Sun, and the diameter is 10.2% of the Sun (88,000 miles).

The closest stellar neighbors are Altair (see #78) at 3.7 lightyears, making it a good two magnitudes brighter than it is in our sky, and 70 Ophiuchi (see #76) at 6.1 lightyears, and also much brighter than we see it.

GJ 752 is located at Right Ascension 19 hours 16 minutes 55 seconds, Declination +05 degrees 10' 08".

104 to 107. *GJ 570* is located in the constellation Libra. It is a complex grouping of three stars and a brown dwarf, 19.3 lightyears away.

Star A is classified as K4.5V, with a mass 76% that of the Sun, and a diameter 77% of the Sun's (716,000 miles). The luminosity is 15.6% of the Sun, and it rotates in 48.3 days. The apparent magnitude is 5.64 (barely visible to the naked eye under very good conditions), and the absolute magnitude is 6.80. The life zone is centered around 0.70 AU from the star (about 67 million miles), but no planets have been detected.

Star B is classified as M1V, with an apparent magnitude of 8.30 and an absolute magnitude of 9.46. Its mass is 55% of the Sun's mass, its diameter is 60% of the Sun's (518,000 miles), and the luminosity is 1.5% of the Sun. It is about 125 AU from Star A (11.6 billion miles). At this distance Star A would be visible but rather dim from a planet around B, while B would be visible only with optical aid from a planet around A.

Star C is classified as M3V, with a mass 35% of the Sun and a luminosity just 0.3% of the Sun. Its apparent magnitude is 9.96, and the absolute magnitude is 11.12. It is less than five AU from star B, and would be brighter than the Full Moon is to us from a planet around star B.

Finally, the last object in this group is a brown dwarf classified as T7.5V. It has a mass about thirty times that of Jupiter, with a surface temperature of 770K (900F). It is 1500 AU from the three stars in this system (about 137 billion miles). At this distance anyone on a planet around any of the stars would have a hard time discovering it.

GJ 570 is located at Right Ascension 14 hours 57 minutes 28 seconds, Declination -21 degrees 24 ' 56".

108. *GJ 754* is a red dwarf, M4V, in the constellation of Telescopium. The apparent magnitude is 12.23, with an absolute magnitude of 13.37. The mass is 16% of the Sun. A RECONS team of Weh-Chun Jao, Todd Henry and others in 2004 found this star was a solar neighbor. It is also listed as *LHS 60*.

GJ 754 is 19.26 lightyears from us. GJ 754 is located at Right Ascension 19 hours 20 minutes 48 seconds, Declination -45 degrees, 33' 30".

109. *GJ 588* is a M2.5V red dwarf in the constellation of Lupus. With a mass 46% of the Sun's mass, it has an apparent magnitude of 9.31, and an absolute magnitude of 10.44. GJ 588 is 19.3 lightyears away.

GJ 588 is located at Right Ascension 15 hours 32 minutes 13 seconds, Declination -41 degrees 16' 32".

110 & 111. *GJ 1005* is a pair of red dwarfs in Cetus. Star A is spectral class M3.5V, with an apparent magnitude of 11.60, and an absolute magnitude of 12.73. Its mass is 18% of the Sun.

Star B is smaller, with a mass only 11% of the Sun. It is M8V, with an apparent magnitude of 14.02 and an absolute magnitude of 15.15. This pair of stars is 19.34 lightyears from us.

GJ 1005 is located at Right Acension 00 hours 15 minutes 28 seconds, Declination -16 degrees 08' 02".

112 & 113. *Achird,* or *Eta Cassiopeiae,* or *GJ 34,* is a binary in the constellation of Cassiopeia. William Herschel discovered this star was a binary in 1779, two years before he became famous by discovering the planet Uranus.

Star A is a G3V star, not dissimilar to the Sun. Its apparent magnitude is 3.46, and absolute magnitude is 4.89. Its diameter is 98% that of the Sun (847,000 miles), and the surface temperature is 5730K. The mass is actually slightly greater than the Sun's. Achird is 19.4 lightyears away.

Star B is spectral class K7V, with an apparent magnitude of 7.21 and an absolute magnitude of 8.34. The diameter is 65% of the Sun, and the mass is 56% of the Sun. Surface temperature is 4100K. Star B is in an elliptical orbit around A that takes 480 years to complete, ranging from 36 AU to 107 AU from one another. With such a separation planets could exist around each, although none have so far been found.

The closest star to Achird is BD +56 degrees 2966 at a distance of 4.9 lightyears, followed by Mu Cassiopeiae at 5.4 lightyears. There are eleven stars within ten lightyears.

Achird is located at Right Ascension 00 hours 49 minutes 06 seconds, Declination +57 degrees 48' 55".

WISE discovered the stars listed here as 13, 58, and 95.

CONSTELLATIONS WITH NEIGHBORING STARS

The following list includes all the constellations, and indicates in which constellation stars in the above list are located.

Andromeda 14, 27, 28
Antlia 45
Apus none
Aquarius 1, 18, 19, 20, 64
Aquila 81, 102, 103
Ara 57
Aries 1, 38, 54
Auriga 91
Bootes none
Caelum none
Camelopardalis 87, 89, 90
Cancer 1, 32, 82, 83
Canes Venatici none
Canis Major 8, 9
Canis Minor 21, 22, 37
Capricorn 1, 95
Carina 65
Cassiopeia 112, 113
Centaurus 2, 3, 4
Cepheus 43, 44
Cetus 10, 17, 33, 35, 63, 95, 110, 111
Chamaeleon none
Circinus none
Columba none
Coma Berenices none

Corona Australis none
Corona Borealis none
Corvus none
Crater none
Crux none
Cygnus 23, 24, 59, 60, 61
Delphinus none
Dorado none
Draco 25, 26, 54, 97
Equuleus none
Eridanus 15, 53, 74, 75, 76, 77, 86, 97
Fornax 72
Gemini 1
Grus 71
Hercules 58, 88
Horologium 34
Hydra none
Hydrus none
Indus 29, 30, 31
Lacerta 77
Leo 1, 6, 69
Leo Minor none
Lepus 96, 98, 99
Libra 1, 104, 105, 106, 107
Lupus 109
Lynx none
Lyra 13, 92
Mensa none
Microscopium 42
Monoceros 36, 46, 47
Musca 62
Norma none
Octans none
Ophiuchus 1, 5, 48, 79, 80
Orion 84, 95, 100
Pavo 39, 40, 101
Pegasus none
Perseus none
Phoenix none
Pictor 41
Pisces 1, 52
Piscis Austrinus 16
Puppis 86
Pyxis none

Reticulum none
Sagitta none
Sagittarius 1, 12
Scorpius 1
Sculptor 49
Scutum none
Serpens none
Sextans 56
Taurus 1
Telescopium 78, 108
Triangulum none
Triangulum Australe none
Tucana none
Ursa Major 7, 66, 67, 68
Ursa Minor none
Vela none
Virgo 1, 17, 50, 51
Volans none
Vulpecula none

ALPHABETICAL LIST OF CLOSEST STARS

The following lists the nearest stars alphabetically, with the star's number in the list above.

A

Achird A	112
Achird B	113
AD Leonis	69
Alpha Centauri A	3
Alpha Centauri B	4
Alpha Centauri C	2
Alsafi	97
Altair	81
AOe 17415.6	55
AX Microscopii	42

B

Barnard's Star	5
BD +68^946	55
Bessel's Star A	23
Bessel's Star B	24
BL Ceti	10

C

CD -46^11540	57
CD -49^13515	71

D

DEN 0255-4700	73
DEN 0817-6155	70
DEN 1048-3956	45
DO Cephei	44
DX Cancri	32
DY Eridani	76

E

EI Cancri	82
Epsilon Eridani	15
Epsilon Indi A	29
Epsilon Indi Ba	30
Epsilon Indi Bb	31
Eta Cassiopeiae A	112
Eta Cassiopeiae B	113
EV Lacertae	77
EZ Aquarii A	18
EZ Aquarii B	20
EZ Aquarii C	19

F

FL Virginis	51

G

G 099-049	84
Gliese 1	49
Gliese 380	68
Gliese 866 A	18
Gliese 866 B	20
Gliese 866 C	19
GJ 34 A	112
GJ 34 B	113
GJ 169 A	89
GJ 169 B	90
GJ 229 A	98
GJ 229 B	99
GJ 251	91
GJ 412 A	66
GJ 412 B	67

GJ 440	62
GJ 445	87
GJ 451	2
GJ 473	50
GJ 570 A	104
GJ 570 B	105
GJ 570 C	106
GJ 570 D	107
GJ 588	108
GJ 674	58
GJ 682	78
GJ 687	55
GJ 693	101
GJ 752 A	102
GJ 752 B	103
GJ 754	108
GJ 832	72
GJ 866 A	18
GJ 866 B	19
GJ 866 C	20
GJ 873	78
GJ 1002	63
GJ 1061	34
GJ 1116 A	82
GJ 1116 B	83
GJ 1245 A	59
GJ 1245 B	60
GJ 1245 C	61
GQ Andromedae	28
Groombridge 34 A	27
Groombridge 34 B	28
Groombridge 1618	68
GX Andromedae	27

H

HH Andromedae	14
HIP 15689	53
HIP 29295	98
HIP 49908	68
HIP 57367	62
HIP 70890	2
HIP 85523	57
HIP 86162	55

K

Kapteyn's Star	41
Keid A	74
Keid B	75
Keid C	76
Krueger 60 A	43
Krueger 60 B	44

L

Lacaille 8760	42
Lacaille 9352	16
Lalande 21185	7
LHS 43	44
LHS 60	108
LHS 288	65
LHS 292	56
LHS 333	50
LHS 449	57
LHS 450	55
LHS 1565	34
LHS 1723	85
LP 656-38	75
LP 816-060	94
LP 944-020	72
Luyten's Star	37
Luyten 726-8 A	10
Luyten 726-8 B	11
Luyten 789-6 A	18
Luyten 789-6 B	20
Luyten 789-6 C	19

O

Omicron-2 Eridani A	74
Omicron-2 Eridani B	75
Omicron-2 Eridani C	76

P

Procyon A	21
Procyon B	22
Proxima Centauri	2
(The) Pup	9

R

Ross 47	100
Ross 128	17
Ross 154	12
Ross 248	14
Ross 614 A	46
Ross 614 B	47
Ross 652	102
Ross 780	64

S

SCR 1845-6357 A	39
SCR 1845-6357 B	40
Sigma Draconis	97
Sirius A	8
Sirius B	9
SO 02353+1652	38
Struve 2398 A	25
Struve 2398 B	26
Sun	1

T

Tau Ceti	33
Teegarden's Star	38
TZ Arietis	54

U

UGPS 0722-05	36
UV Ceti	11

V

Van Biesbroeck's Star	103
Van Maanen's Star	52
V 645 Centauri	2
V 1581 Cygni A	59
V 1581 Cygni B	60
V 1581 Cygni C	61

W

WISE 1541 -2250	13
WISE J1741 +2553	58
WISE J0254 +0223	95
Wolf 359	6
Wolf 424 A	50
Wolf 424 B	51
Wolf 498	88
Wolf 1055	102
Wolf 1061	48
Wolf 1453	93
WX Ursae Majoris	67

Y

YZ Ceti	35

NUMBERS

2MA 0415-0935	96
2MA 0939-2448	86
2MA 1835+3259	92
40 Eridani A	74
40 Eridani B	75
40 Eridani C	76
61 Cygni A	23
61 Cygni B	24
70 Ophiuchi A	79
70 Ophiuchi B	80

CLASSES OF STARS

A Stars	8, 81
Brown Dwarfs	13*, 30, 31, 36, 40, 58, 70, 72, 73#, 86, 95, 96, 99, 107 *Class Y. #Class L. All others are Classified T. (77 may have an unconfirmed brown dwarf companion.)
F Stars	21
G Stars	1, 3, 33, 97
K Stars	4, 15, 23, 24, 29, 68, 74, 79, 80, 93?, 104
M0 to M2	7, 16, 27, 41, 42, 49, 66, 71, 88, 93?, 94, 98, 105
M3 to M4	5, 12, 17, 25, 26, 28, 35, 37, 43, 44, 46, 48, 53, 55, 57, 64, 69, 75, 77, 78, 84, 87, 89, 91, 100, 101, 102, 106, 108, 110
M5 to M6	2, 6, 10, 11, 14, 18, 19, 20, 32, 34, 38, 50, 53, 56, 59, 60, 63, 65, 67, 77, 83, 85, 90, 111
M7 to M9	39, 45, 47, 51, 61, 92, 103
Planets	1, (5), (7), 15, (23), (29), 33, 57, 64, 65, (68), 71, (79), (80), (102) ( ) indicates unconfirmed, suspected, or challenged
White Dwarfs	9, 22, 52, 62, 75, 90
Binaries	8-9, 10-11, 21-22, 23-24, 25-26, 27-28, 39-40, 43-44, 46-47, 50-51, 66-67, 79-80, 89-90, 98-99, 110-111, 112-113
Triplets	2-4, 18-20, 29-31, 59-61, 74-76
Fourplex	104-107

SOURCES

NASA

Norton's Star Atlas

RECONS: The Research Consortium on Nearby Stars, based at Georgia State University

SIMBAD: Strasbourg, France

Solstation

NASA Jet Propulsion Laboratory Small Body Database
http://ssd.jpl.nasa.gov/sbdb.cgi?sstr=4897

Hamilton Planetarium Scholarship Fund
http://www.planetariumscholars.webs.com

INDEX OF PEOPLE

Please note that people below are indexed by the number of the star(s) where they are mentioned. If they are not mentioned with any star, the page number where they are mentioned is given.

CPSIA information can be obtained at www.ICGtesting.com
Printed in the USA
BVOW051143230412

288360BV00004B/1/P

9 781618 971326